ROUTLEDGE LIBRARY EDITIONS:
THE ECONOMICS AND BUSINESS OF TECHNOLOGY

Volume 32

THE COMPUTER AND THE CLERK

THE COMPUTER AND THE CLERK

ENID MUMFORD AND OLIVE BANKS

Routledge
Taylor & Francis Group
LONDON AND NEW YORK

First published in 1967 by Routledge & Kegan Paul

This edition first published in 2018
by Routledge
4 Park Square, Milton Park, Abingdon, Oxon OX14 4RN
605 Third Avenue, New York, NY 10017

Routledge is an imprint of the Taylor & Francis Group, an informa business

British Library Cataloguing in Publication Data
A catalogue record for this book is available from the British Library

ISBN: 978-1-138-50336-6 (Set)
ISBN: 978-1-351-06690-7 (Set) (ebk)
ISBN: 978-0-8153-8720-6 (Volume 32) (hbk)
ISBN: 978-1-351-17368-1 (Volume 32) (ebk)

Publisher's Note
The publisher has gone to great lengths to ensure the quality of this reprint but points out that some imperfections in the original copies may be apparent.

Disclaimer
The publisher has made every effort to trace copyright holders and would welcome correspondence from those they have been unable to trace.

THE COMPUTER AND THE CLERK

by

ENID MUMFORD AND OLIVE BANKS

LONDON

ROUTLEDGE AND KEGAN PAUL

First published 1967
Routledge & Kegan Paul Limited
Broadway House, 68–74, Carter Lane
London, E.C.4

Printed in Great Britain
at The Villafield Press
Glasgow

CONTENTS

TABLES

ACKNOWLEDGEMENTS

The authors would like to express their sincere gratitude to the two firms which participated in this study. They were prepared to tolerate research at a time when they were both struggling to introduce a new and untried system of work. Yet they bore with us patiently and were always anxious to co-operate and to provide us with the information we sought.

We must also thank our colleagues on the staff of the Liverpool University Department of Social Science who helped us with the field work and with the analysis of research data. In particular Mr. R. Blackburn whose statistical advice was invaluable and Mrs. I. Hordley and Dr. K. Prandy who both participated in the interview programme and helped prepare some of the statistical tables.

Throughout the final stages of the report writing we have benefited greatly from the suggestions, help and criticisms of the editor, Professor M. Fogarty. We have also profited from the comments and criticisms of Professor Lord Simey, Mr. J. A. Banks, Dr. R. Cooper and Mr. T. B. Ward. We would thank all these for their kindness in reading drafts of this manuscript.

Lastly, we would thank the Department of Industrial and Scientific Research for making the project possible, the Social Science Department typing staff for dealing with the manuscript so expeditiously and Mr. G. Wilson for preparing the diagrams.

INTRODUCTION

How the study came about

The Department of Social Science at Liverpool University has always been interested in problems of technology and change and some years ago co-operated with a number of other European countries in studying the attitudes of workers in the steel industry to changes arising out of new production systems.[1] An international committee for social research in industry was formed as a consequence of this liaison, with the objective of building up a body of sociological knowledge concerning the effects of new techniques on the formal and informal structures of firms and on different occupational groups within these firms. Although appreciating that the impact of change would vary from one organisation to another according to its nature and extent, its location and how rapidly it was introduced, the committee believed that some generalisations about attitudes to change and its social and organisational consequences could be formulated, given data from a number of comparable firms in different countries.

At a meeting of the committee held in Paris in 1957 it was decided that research into the effects of automation on the white-collar worker would be of value in view of the speed with which technological developments were affecting the office and the paucity of information about white-collar workers as a group. Although the project, as planned, was primarily directed at obtaining information on the results of change; the international nature of the research was seen as providing an opportunity for securing comparative data of a more theoretical kind. It was therefore decided to include some study of social stratification, considering, for example, the social class background of white-collar workers and their social aspirations for themselves and for their children.

[1] For a report of the British Study see, W. H. Scott, J. A. Banks, A. H. Halsey and T. Lupton, *Technical Change and Industrial Relations* (Liverpool University Press, 1956), and O. Banks, *The Attitudes of Steel Workers to Technical Change* (Liverpool University Press, 1960).

Each national team submitted a research outline to the sponsoring organisation in its own country. In Britain this was the Department of Scientific and Industrial Research. At this time a considerable number of countries were anxious to participate in the research although, for a variety of reasons, the only ones to complete their studies were Britain, France, Germany and Sweden.

The international committee appreciated that not all countries would have identical interests in pursuing the study and that, while it was essential that certain areas of investigation should be covered by all participating research bodies, the project must also be made flexible enough to allow each team to pursue its own particular interests. The final research design, therefore, focused on four areas as a minimum requirement. These were (1) the introduction and handling of the change. (2) Changes in the occupational structure of the labour force as a result of automation. (3) Changes in the social structure—that is in the relationships of one group with another, and (4) labour relations including consultation, communication and management–union relations.

Each country was asked to undertake a study of at least one, but preferably two, large industrial or commercial organisations. Government bodies affected by automation were to be excluded from the survey as it was thought that rather special conditions prevailed in these which would make them less valuable for purposes of comparison. Ideally, the project was to take the form of an on-going study of change. That is, the various firms were to be observed before the change took place, during the transition period from one system to another, and after the change had been completed and the system had settled down. Britain, France and Sweden were able to meet this requirement but Germany was pressed by its sponsor for speedy research results and so had to confine itself to a post-change study only.[2]

The British project

In this country the research was started in 1957 although, through an unforeseen setback, little progress was made until 1960.

[2] The summaries of the international results are to be found in W. H. Scott, *Office Automation and the Non-Manual Worker* (Paris, O.E.C.D., 1962), and W. H. Scott, *Office Automation: Administrative and human problems* (Paris, O.E.C.D., 1965).

We planned to attempt a three-phase study of two firms, one a business house, the other a manufacturing company. Phase One was to be an attitude survey of all clerks in departments likely to be affected by the computer, together with unstructured interviews with key persons, analysis of documentary sources and observations of departmental organisation and work flow before the change-over took place. Phase Two would be a study of the mechanisms of change, what happened and what problems were encountered during the period between the computer's arrival and the eventual transfer of work. Phase Three would be a second attitude survey and a re-examination of departmental organisation and work flow. This last stage was scheduled to take place six months after the computer had taken over, when, it was hoped, stability would have been restored to the firm.

The pre-change attitude survey was designed to obtain information on the clerk's social class origin, educational background, formal qualifications, occupational history and present job; together with attitude data on working conditions, promotion prospects, the position of clerks *vis-à-vis* other occupational groups, trade union membership and the likely effects of automation on the role and position of the clerk. The post-change survey would cover broadly the same areas but would concentrate on eliciting differences in the clerk's perceptions of his job, role and status after office automation compared with before.

Our first operational problem was to find a firm which was planning to introduce a computer and which would permit an investigation of the change-over to an automated system of work. This was not easy. Computers were still relatively scarce in this country—a survey made at the end of 1958 showed that the five major manufacturers had delivered one hundred up to that date. However, the Department of Social Science had been carrying out some research into a different problem in A.O.C., the local subsidiary of a large manufacturing combine, and this firm announced that it was proposing to replace its existing Hollerith punch card equipment with a computer.[3] It was asked to participate in the office automation research and agreed to do so. A.O.C. is of course a fictitious name adopted to conceal the

[3] See J. A. Banks, *Industrial Participation* (Liverpool University Press, 1963).

company's identity and we shall follow this practice when we discuss other firms.

In 1957 a preliminary attitude survey on the lines outlined earlier was carried out in this firm and 114 clerks were interviewed out of a total of 128. At the same time background information was collected, personnel records were examined, and information obtained on how the change was to be introduced. By the end of 1957 a reasonably comprehensive picture of the pre-change situation had been revealed. Unfortunately, there now occurred the first of a series of difficulties that were to dog the research team throughout the project. The computer arrived but was found unsuitable for the firm's requirements. Eight months after its installation, when only a small amount of work had been transferred, it was sent back to the manufacturers and a decision was taken not to replace it. This meant that our research in A.O.C. had to be abandoned.

At the beginning of 1960 we learned that a bank was proposing to install a computer to take over routine branch accountancy. This bank, which we will call Royal Exchange, was prepared to accept the research and interviewing commenced there in August 1960. Initially the bank planned to transfer the work of nine local branches onto the computer but this number was revised to five soon after our investigations began. It took the research team some time to obtain permission to interview clerks in a number of these branches and so the pre-change phase of the survey was not finally completed until April 1962. Despite these delays, interviewing was over well in advance of the computer's encroachment on the five office situations.

Interviewing for the post-change attitude survey was commenced in April 1963 and completed in four branches by June 1964. The bank now decided to abandon its plans to transfer the work of the fifth and largest branch to the computer. This was an unforeseen setback as a great deal of time had been spent in securing pre-change data in this branch. However, we decided to complete the interviewing programme in the hope of securing some interesting comparative data. For example, it would be possible to establish how clerks in a non-affected branch viewed office automation. These final interviews took place at the beginning of 1965.

Simultaneously with the interviewing programme, discussions

took place with branch managers and other senior staff and with the manager of the bank's Organisation and Methods Department. Observations were made in the computer centre and many informal conversations held with the programming staff there. Contact was also made with the bank's Staff Association and with the National Union of Bank Employees.

Because of the abrupt termination of the project in A.O.C. we considered it essential to find a second firm for study and an approach was made to Carters, a local manufacturing firm which the research team had heard was proposing to install a computer. This company agreed to provide research facilities and by December 1960 the pre-change attitude survey had been completed there. Carters also proved extremely slow in transferring work to its computer and the post-change survey could not be completed until October 1963. Even by this date only a part of the work originally planned for transfer to electronic data processing (E.D.P.) had in fact been taken over. The post-change survey could not be postponed any longer, however, as the firm had now decided to scrap its first computer and replace it with a larger and more reliable model.

Dilemmas and objectives

Industrial sociologists are prone to try to write their books for two types of readers simultaneously, their academic colleagues and practising managers. In attempting to meet the needs of these very different groups at one and the same time they often end up by pleasing nobody. The academics complain that the analysis of data has not been shown to be sufficiently rigorous while the managers say that the detailed exposition of research material makes the book hard for them to digest—they only require general principles. In writing up the research project described in the following pages we have encountered this dilemma and it has not been an easy one to resolve. However, as *The Computer and the Clerk* is being published in a Business Studies Series we have tried to lean more towards the managers needs and, while still setting out our evidence, doing this as simply and clearly as possible and drawing practical conclusions from it. We hope the book will be of value to those concerned with introducing technical change and we would make the point most forcibly that, although we were studying the introduction of computers, the kinds of problems emerging from our study might be found with many kinds of technical change. Similarly the analytical approach we set out in the last chapters would, we suggest, be of value in any change situation. One further point can usefully be emphasised here. The fact that our case study firms were to some extent 'pioneers' in their use of computers for business purposes meant that the change process proved particularly stressful and difficult. Today computers are technically much more reliable and can be introduced without the delays and breakdowns that we shall be describing in subsequent chapters. Yet this does not mean that human relations problems will be any easier to handle. On the contrary it seems probable that as computers are used to control larger areas of business activity so the social and organisational problems associated with their assimilation will become increasingly complex.

PART I

Offices in transition

CHAPTER 1

Some problems of office automation

Technical change and its problems have been extensively studied by sociologists and others for many years past but, until recently, an examination of the dynamics and consequences of this kind of change has been confined almost entirely to the factory shop floor. We now have studies of technical change and its impact on production techniques, work organisation and human relations in such industries as car manufacture, steel and coal, to mention but a few.[1]

Today, however, with the advent of computers and electronic data processing a quite new kind of technical change is becoming commonplace—office automation. This affects clerical rather than production departments, and clerks instead of manual workers; it is to be found in purely commercial and administrative organisations such as banks, insurance companies and local authorities as well as in the clerical services of manufacturing firms.

Technical change in the office may be more difficult than on the shop floor

The novelty of this kind of major technical change in offices leads immediately to a number of difficulties which may not occur or may be more easily overcome when technical change is introduced on the shop floor.

First, the production departments of rapidly developing manufacturing firms have become accustomed to handling and

[1] See, for example, C. R. Walker, and R. Guest, *Man on the Assembly Line* (Harvard University Press, 1952). W. H. Scott, J. A. Banks, A. H. Halsey, and T. Lupton, *Technical Change and Industrial Relations* (Liverpool University Press, 1956), and E. L. Trist, G. W. Higgin, H. Murray and A. B. Pollock, *Organisational Choice* (Tavistock Publications, 1963.)

adjusting to frequent changes in technology and many have learned how to do this efficiently and easily. Offices, in contrast, have been used to a much slower rate of change. While the majority today are at least partially mechanised through the introduction of equipment such as typewriters, adding and accounting machines and punched card systems, these developments have come about gradually and caused few problems of assimilation and adjustment. Office automation has now altered this slow moving absorption of innovation. The introduction of a computer can produce quite dramatic changes in an office situation; for example, in the organisation of clerical, information and control functions, in the kind of work office staff are required to do, and in the number of clerks and managers needed to do this. Often these consequences are not anticipated by management involved in the change; they are taken by surprise when what they envisaged as a simple change of system acts like a bomb explosion and causes repercussions that spread through many departments.

Second, not only are production departments more likely to be experienced in handling technical change, management there is usually under some pressure to proceed thoughtfully and carefully. This is because workers affected by the change will almost certainly be union members and they will expect their union officials to be in continual negotiation with management to safeguard their interests. With the exception of local government workers, who are highly unionised, and bank clerks who are likely to belong to either bank staff associations or N.U.B.E., most clerks are unlikely to have this kind of watchdog. They are therefore less protected than the shop floor worker and more at the mercy of the vagaries of their bosses. At the same time, managements which are free from union pressure may be tempted to move rapidly ahead with technical change without thinking through the human relations implications of what they are doing.

Third, whereas change on the shop floor is handled by production managers, engineers and other technical specialists who are well used to this kind of re-organisation, in an office installing a computer for the first time, responsibility for introducing the machine may be given to the firm's accountant or, if the firm has one, to the organisation and methods department. The former may not be used to handling change at all, the latter may regard a computer as a simple move from one system of work to another.

Neither are likely to anticipate or understand the far-reaching organisational and human relations consequences of what they are doing. Again and again in our research we came across examples of firms running into difficulties because the social effects of office automation were not foreseen or planned for, and we found managements were astonished and hurt when their employees refused to co-operate or started looking for other jobs.

Fourth, the shop floor worker is himself more accustomed to change than the clerk and is, perhaps, less likely to be disturbed by it. There has always been some degree of insecurity attached to his job and he has adjusted to this. Not so the office worker who may have worked for years in a secure and sheltered environment and who expects this condition to continue. He receives a rude shock when he finds that a piece of electronic equipment is about to take over his work and he is told that a machine can complete clerical operations more accurately and efficiently than he ever could.

Consequently computers present a considerable challenge to those managers who have to handle their introduction and to those clerks and managers who experience their effects.

Our approach

There is a great need for systematic information on what happens when a firm installs a computer, on the kinds of problems that are likely to be encountered and on the kinds of strategy that must be adopted in order to avoid these. At the moment this information is patchy and one-sided. There are many articles and books on the technical aspects of computer introduction, virtually nothing on the organisational impact of the extensive use of computers involved in integrated data processing—although industry must adopt an integrated approach if it is to use computers effectively—and very little on the social impact of computers on the clerical labour force.

This book attempts in a small way to provide some systematic information on the social aspects of computer installation by describing how two firms introduced computers and the problems they encountered in doing so. Our data is presented as follows:

Chapter 2 provides some background information on clerks and office automation and sets out the hypotheses that our research is designed to test. Chapters 3-6 will present the situation in our first firm, the Royal Exchange Bank, describing the sequence of events associated with the introduction of the computer and analysing the interview data which we obtained from staff in four automated branches and in one branch which underwent only a very slight change. In Chapters 7 and 8 the same outline is followed for our second firm, Carters. Chapter 9 examines the consequences of the change for the two firms and the clerks working there and compares our research findings with similar studies in other countries. Chapter 10 attempts to identify those variables which influence clerical attitudes to change and, from this analysis, we will try to arrive at some general conclusions concerning the management of change. Chapter 11 presents a strategy for change which, it is hoped, will provide managers with some pointers on how to avoid stress, anxiety and opposition to change from developing in their employees.

Some introductory points

The impact a computer has on a firm will depend very much on what it is used for. Firms which merely automate existing clerical procedures and use their computers as high powered accounting machines may find that no fundamental organisational changes are caused by the new system. In contrast, firms which introduce comprehensive integrated data processing systems and use their machines to rationalise various business functions are likely to find that the structure of their company has been dramatically altered. Departments may be eliminated or combined and the content of both managerial and clerical jobs will be greatly changed. But, however a computer is used, and in Britain so far the practice has been to adopt the first approach, the move from one system of work to another must lead to the abandonment of old methods, practices and ideas and to the adoption of new ones. These new procedures alter the traditional office situation; they will be seen by office workers as affecting their present jobs and their future careers, and both clerks and managers are likely to react to them in a very ambivalent way.

Staffing problems

The first and crucial problem for the clerk is that most computers are introduced to cut labour costs. They therefore have the potential of causing redundancy and throwing the clerk out of work. The clerk knows this and despite assurances from his employers that this will not happen, he is usually a worried man. This worry is not because he sees his colleagues in other firms losing their jobs—for the amount of redundancy caused by office automation in this country is negligible—but because he is now subjected to a great deal of propaganda about machines taking over his work. There are jokes in *Punch*, television programmes showing offices without clerks, warnings from white-collar unions, articles in almost every journal he picks up, and so on.

This current nervousness about job security means that any firm contemplating a computer must find out at an early stage exactly how and to what extent the machine is going to affect its labour force. It must give reassurances about redundancy as soon as the machine is ordered and, if some reduction in staff numbers is likely to occur, it must work out comprehensive advance schemes for running down the labour force through natural wastage. Fortunately the majority of clerks affected by a computer are usually young women. Women today get married at an early age and there is a high labour turnover in this group. This means that the problem of reducing staff numbers is not a difficult one for management provided plans are made and implemented well before the computer arrives.

Even though redundancy can be avoided, a firm is still left with the problem of staff displacement. Unless work is expanding at a proportionate rate, the introduction of a computer must reduce the numbers of staff required by an affected department, and this will mean that plans will have to be made for transferring staff elsewhere in the organisation. This presents its own problems; suitable jobs must be identified, re-training will be necessary and the firm may have to make special provision for this, while the staff themselves may object to being moved. Firms seem to have little trouble with their young clerks who are adaptable and have, as yet, little job investment. Elderly clerks can often be retired early, perhaps with a golden handshake. The difficulty lies more with the middle-aged clerk or departmental supervisor who can have spent twenty years in the same job. He may find it very difficult indeed to pick up new knowledge and skills at this late stage in his career. Re-training, particularly of the older worker, seems a problem warranting far more investigation than it is receiving at present.

Those staff remaining in departments taken over by the computer will, of course, be considerably affected by the new system and there may be changes in the content, skill requirements and status of their jobs. Some clerks, for example, will be transformed into machine operators punching tape or cards for the computer. In a later chapter we will show how girls in Royal Exchange Bank were changed from accounting machine operators to punched tape machinists. They believed this new job to be less skilled and more monotonous than their previous one. They also complained of a lowering of job status and compared themselves with some bitterness to factory hands. This last reaction was a subjective and emotional one, as in the bank's eyes their status had not altered at all.

Departmental supervisors trained in traditional office routines are likely to find the new technology hard to understand; they may find their functions reduced in range and variety and they may have considerably fewer staff to control. It can prove difficult to convince this group that an automated system has any advantage over traditional methods.

Staff reactions

Our research findings will show that our two case study firms—and we found this in other firms visited during the course of the investigation—had difficulty in introducing the kind of rapid technical change associated with electronic data processing. They made mistakes in their handling of the change process and in the policies they adopted, and because of these they evoked anxious and hostile reactions in some of their staff. There would, of course, have been some fear and antagonism even if the change-over had been handled perfectly, but imperfect strategies for change did seem to magnify the problem.

Anxiety was not confined to clerks but occurred at every level in the firm. Middle managers saw their status and autonomy threatened, junior management resented the rise of the new specialist computer groups, while clerks were anxious about their jobs, their security, and the nature of their new work. Although Royal Exchange was able to secure the co-operation of its staff, other firms were less fortunate and we came across instances where clerks believed that in accepting a computer they were 'cutting their own throats'. Both our firms suffered greatly from the production of corrupt input data in the early days. In Royal Exchange this was largely a technical problem due to machine and tape defects, but in Carters there was some evidence that clerks in affected departments were not motivated to produce accurate information.

Why was there anxiety and antagonism about office automation? We identified a number of reasons for this. The first was management's unawareness of the social implications of the change they were introducing. In one firm the principal innovator was an organisation and methods expert, in the other, an accountant. This meant that electronic data processing in both companies was seen primarily as a systems change and its consequences for human relations were either ignored or not understood.

The second was that the two managements were defective in their communication and consultation policies. One firm did not wish to let staff know what was going to happen, the other tried to inform staff but was only partially successful in doing so. Rumour was everywhere given as the principal source of information, yet Royal Exchange management was trying hard to keep its staff informed of what was taking place and we found that those clerks who recalled hearing about the computer from official sources were more likely to react approvingly than those who said they had heard unofficially.

The third was that the pressures of the change period were themselves a source of stress and anxiety. In both firms the preliminary time estimates for the completion of the changeover proved wildly unrealistic. Computer staff became worried because they could not meet deadlines and were subjected to pressure from senior management who believed they should be able to. Computer breakdown caused plans to fall behind schedule and a backlog of work to accumulate. This in turn led to continual overtime work in the evenings and at weekends. Both computer personnel and clerical staff in affected departments suffered from a tremendous increase in work during the change-over period and this caused physical and mental stress. Staff became tired and with certain individuals this led to a drop in morale.

Our investigations led us to the conclusion that the introduction of a computer can only be carried through in a calm and ordered way if the changeover is preceded by meticulous planning. In our final chapters we discuss the different factors which we believe influence employee attitudes to change and which therefore must be taken into account if successful plans are to be made.

Our main conclusions

We hope that this study will bring out three main points of importance to the practical manager.

1 That introducing a computer is not merely a simple change from one system of work to another. It is something much more fundamental and far reaching, which can affect the structure of a firm, the organisation of departments, the nature of work and the attitudes and behaviour of employees. Technical change is like a bomb explosion. If no precautions are taken there may be widespread panic and confusion. However, if the area of impact is known in advance with some precision, precautions can be taken that will do much to cushion the shock.

2 That office automation has quite different implications for different individuals and groups within a firm and in consequence, people will react to it in different ways according to their own particular needs, values and interests. Some will welcome a computer, others will be neutral, and others again will be hostile or afraid, seeing the new system as a personal threat. It follows that because people react differently they will need to be treated differently and any blanket policy for change is likely to have only a limited success.

3 That the introduction of any major technical change must involve the formulation of policies and strategies which are based on a careful preliminary analysis of all the variables operating in the change situation. This is perhaps our most fundamental point. We believe that adequate prediction, planning, and the successful execution of plans, can only be achieved if a situation is looked at as a totality and analysed in this way. In order to do this, a manage-

ment must take into account technical, organisational and human relations factors.

At present British management seems prone to concentrate on technical factors and to ignore or fail to understand the organisational or social consequences of its innovations. In the computer field, managers, computer manufacturers and consultants often appear content to do a partial analysis of the effects of electronic data processing. Great time and attention is devoted to the systems aspects of the change and to identifying likely technical problems. Virtually no attention may be given to the sociological problems of seeing how, for example, a firm's management structure may change once managerial functions are automated and how this may, in turn, affect the balance of power within the firm; or to finding out in advance the likely reactions of the different groups whose work will be affected by the new system.

This incomplete kind of analysis is, of course, due to the kinds of specialists traditionally brought in to advise on or to deal with technical change, for usually these are engineers or systems experts. Certainly this kind of person is indispensable to innovation but it can be suggested that working alone he can only cover some of the problem areas, and indeed he is only qualified to do so. Increasingly in the future we would like to see the sociologist and the social psychologist working side by side with the technical expert, so that together they can examine a situation or a problem from both the social and the technical points of view.

Happily there are signs that this is about to happen. Consultants are now setting up social science units, some computer manufacturers are thinking about comprehensive customer advisory services in which both social and technical problems will be dealt with, and one or two firms are employing sociologists and psychologists in their management services departments. But the use of non-technical experts is still the exception rather than the rule.

We hope that this book, by emphasising the non-technical problems associated with change, will persuade some managements that social science can help them to innovate more painlessly than they do at present.

The two computer installations which we describe and discuss in the following chapters are first generation machines and, because the computer industry is developing at such speed, both would now be regarded as out of date. Readers may be tempted to put down the book at this point, assuming that anything we say can have little relevance to problems today. We would, however, ask them not to do this because we believe that although computers have developed in the last few years, the problems of handling change have not altered accordingly. The human relations problems we discuss in the following chapters are just as likely to happen in a firm installing a computer at present as in our two case study firms. The management of change is not made easier because the equipment involved is technically superior, in fact we would suggest that it can be more difficult. Our two computers suffered a great deal of technical trouble. They both broke down repeatedly in their early days and the firms had to put up with hold-ups and delays. But, from the point of view of achieving staff acceptance of automation, this could have been an advantage. It gave management time—time for clerks to get used to the idea of a new system, time for management to provide cushions for any hardship that might occur, and time to sell the idea of a computer to everyone involved. Neither of our firms used the unscheduled time in this positive way, but they could have done so had they wished.

Today, however, this time bonus is no longer available unless management deliberately slows down the change process. Theoretically, simple work procedures can be transferred to a computer within a few months of its arrival in a firm and the modern machine should not cause delays by breaking down or making mistakes. This means that there will be far less time for staff adjustment and, in consequence, pre-change analysis and planning will need to be done very carefully indeed if staff are not to be subjected to intolerable stress and anxiety. In this country our technical competence is developing at a tremendous pace, our human relations knowledge rather more slowly. But, despite this, a great deal of information on human behaviour is now available if only managers can be persuaded to familiarise themselves with it. Unless our industries become fully automatic and control themselves, managers are going to have to cope with problems of introducing change for a very long time to come.

CHAPTER 2

Clerks and social change

This chapter provides some background material against which our survey material can later be examined. Firstly, it contains a short historical account of those changes in the role and status of the clerk that are likely to account for some of his present day attitudes. Secondly, it gives a brief description of office mechanisation and what it involves, and thirdly it describes some of the processes and problems of social change that are likely to occur once extensive mechanisation is introduced.

Changes in the role and status of the clerk

Lockwood (1958) in *The Blackcoated* provides a description of the condition of the clerk in the last century. By 1851, he suggests, there were around seventy or eighty thousand clerks in this country, almost all male. These clerks worked in small counting houses, maintained close and personal relationships with their employers and identified with the interests of their firms. They had middle-class aspirations, job security and the likelihood of promotion, providing only that they could make themselves so knowledgeable in the business affairs of their employers that they became indispensable. Because of the independent nature of their jobs and their physical separation, as a rule, from large numbers of other clerks, they did not recognise themselves as a homogeneous group with common interests and there was little inducement to band together as a collective force.

During this period the manual workers in the factories were in a very different situation. They were now grouped in large numbers in machine shops with goals and interests that were separate from and often opposed to those of management. Poor conditions, low wages and exploiting entrepreneurs created a working environment that was, as Lockwood points out, 'highly conducive to the emergence of class conscious action', and a stimulus to unionisation. In contrast to this it is only quite recently that clerks, in any numbers, have come to recognise that trade unions and collective action offer means for improving their position.[1]

The clerical function altered slowly and undramatically over time. With increasing commerce and industrialisation, offices grew larger and a new kind of labour appeared, the woman clerk. Sinclair Lewis describes in one of his novels a pre-first world war office of managers, bright young men and . . . 'Somewhat outside the main hierarchy the small corps of private secretaries, each the daily confidante of one of the gods. Nevertheless, these confidantes were not able to "associate" with the gods, or "be friendly", in coat room or rest room or elevator with the *unrecognised horde of girls* who merely copied or took the bright young men's dictation.'[2]

Office workers and, in particular, the numbers of women employed in offices, continued to increase until by 1951 there were over two million clerks, of which six out of ten were female (Lockwood, 1958). This trend still continues; in the international survey of automation and the non-manual worker it was usual to find that about 75% of the office personnel in the firms studied were women. With increasing mechanisation it seems likely that clerical work at the routine level will fall almost exclusively into the hands of girls and women.[3] The term 'white-collar' as applied

[1] This is not to suggest that white-collar unions have only recently come on the scene. A number of white-collar unions were going through their early development stages during the last century. Similarly, manual unions, although earlier in origin, were not widely established until this century. See R. Blackburn 'Unionisation and Union Character in Banking', unpublished Ph.D. thesis, for further information on this point.

[2] This quotation from Sinclair Lewis's book *The Job* appears in *White Collar* by C. Wright Mills (New York: Oxford University Press, 1951).

[3] Jack Stieber of Michigan State University in a talk on 'The Impact of Automation on White-Collar Workers' given to the American Midwest Economics Association, April 12th, 1957, pointed out that in the U.S.A. in 1956, 67% of all clerical workers were women as compared with 53% in 1940. Estimates of labour participation rates made by the U.S. Census Bureau indicate that the proportion of women (14 years and older) in the labour force will increase from 31·3% in 1950 to 34·9% in 1965. During this period the male participation rate is expected to decline from 83·3% to 79·7%.

to the clerks then becomes a misnomer and should be replaced with a phrase more appropriate for a predominantly female group. Lockwood uses 'white bloused'.

Why is so much of clerical work being taken over by women? The answer seems fairly obvious. Women—or rather girls, for the great majority are under twenty-five—will, at present, accept routine jobs that are unlikely to lead anywhere. For most of them matrimony is their principal objective or interest and work is regarded as temporary and incidental, rather than central in their lives. Because of this they are willing to put up with tasks that seem intrinsically dull and for the same reason they are not unduly anxious for promotion. Male clerks, in contrast, tend to be very interested in and anxious about promotion. In addition, girls can easily be trained for office machine jobs requiring a high degree of dexterity and they also represent a pool of cheap labour. Thus, women, even when they are employed on the same work as men, rarely get equal financial rewards and today mechanisation makes it possible to create a new type of work, to designate it as a woman's job and to offer a female rate of pay. Mechanisation is now probably the most important single factor in the 'feminisation' of the routine clerk although there are other influences tending to raise the proportion of female clerks. The growth in the scale of offices, for example, produces of itself a division of labour which separates out the more highly skilled from the less skilled work, which is then suitable for relatively short-term female labour. This passive acceptance by women of low grade, routine work may, of course, not continue in the future. Girls as well as boys are being encouraged to train for skilled work and increasingly women are qualifying for careers before marriage and returning to these once their children start school.

The male clerk appears more and more to be found in managerial posts and in clerical jobs requiring a relatively high degree of skill and initiative. These may be 'traditional' clerical skills, costing, pricing, and so on, or they may be new skills required by modern office technology such as programming and systems analysis. In the future, the few men remaining in clerical jobs will be 'juniors' working their way up and the routine male clerk, as a career grade, will become extinct. As those male clerks still to be found in large numbers in such occupations as local government and banking show signs of becoming increasingly dissatisfied with their position and status at work and in society, before going on to examine in more detail the modern development of mechanisation and its effect on the clerk, it may be useful to consider the extent to which the routine male clerk's position has deteriorated compared with that of the manual worker.

The clerk has always believed that the conditions under which he works, his pay, his prospects, and his relationships with his employer, all entitle him to middle-class status, and he has tried to emphasise this status by sending his children to grammar school, dressing in a conventional middle-class way and living in a house in a middle-class neighbourhood. Today this traditional superiority of non-manual over manual work seems to be weakening and is increasingly being questioned, both by the manual worker and by the clerk himself (Lockwood, 1958). One reason for the lowering of the clerk's status is to be found in changing clerical working conditions. As the ratio of non-manual to manual workers has decreased so offices have become larger and document processing stimulated. The modern office is no longer a small and friendly place. On the contrary it may be enormous, full of machines and like a factory in its flow of work. The advent of large scale mechanisation, and in particular the electronic computer, has increased its resemblance to the shop floor and many jobs, usually those performed by women, make demands on the clerk similar to those made on the worker by factory production lines. Wright Mills (1951) suggests that there are now few, if any, features of wage work (except heavy toil, which is becoming rare) that do not also characterise at least some white-collar work.

The clerk's pay, too, has fallen relative to that of the manual worker. The last war and the post-war situation brought about a considerable narrowing of the gap between manual and non-manual groups. From 1939 onwards wages increased more rapidly than salaries and the average clerk is now roughly on the same income level as the average manual worker, or slightly below (Lockwood, 1958). He is well below the income level of the craftsman. A factor here is, of course, the prevalence of paid overtime for the manual worker. The clerk is unlikely to work regular overtime and when he does this may be accepted as a requirement of his job and he receives no compensation.[4] (Both of

[4] It has been suggested, however, that the income differences between these two groups are less than they appear. R. Hamilton in an article

our two case study firms did make overtime payments to clerks.)

Another equalising factor is the changed approach to fringe benefits in industry, although in this instance it is the manual worker moving up, rather than the white-collar worker moving down. In the past the clerk has had many privileges which have not been awarded to the shop floor. He has had longer holidays, holidays with pay, superannuation, sick pay and a shorter working day. Some years ago it was unusual to find many manual workers in receipt of these. It was believed that the blue-collar man must only be paid for the hours he was at work and the amount he produced. The climate of industrial opinion gradually swung away from this discrimination and industrialists began asking why, for example, the skilled fitter with forty years service in the same firm should receive nothing if he was ill, while his young daughter, only six months in the typing pool, should be given full pay. Today many firms have comprehensive sick benefit and superannuation schemes for works and staff alike and there is a move towards similar hours of work for the two groups during the normal working day.

With promotion there is still a difference between manual and white-collar workers, this time to the advantage of the clerk. The employment of women in routine clerical jobs and their lack of interest in career prospects has meant increased promotion opportunities for men. Lockwood refers to a study of labour mobility in Great Britain between 1945 and 1949 which shows that of those who began their occupational lives as clerks, one-third moved up the occupation ladder and became managers, one-third stayed as clerks, one-third moved down to manual work.[5] He suspects, however, that this mobility upwards will diminish as industry increasingly recognises the need for formal management training.

Because the working conditions, pay, fringe benefits and status

in the *British Journal of Sociology*, XIV, 4th December, 1963, 'The Income Difference Between Skilled and White Collar Workers' says that statistics are biased by the predominance of women in the white-collar groups, by the inclusion of well paid non-manual groups such as foremen in the manual worker category, and by no account being taken of career patterns.

[5] This period is, of course, atypical in that industry was short of managers at the end of the war.

of clerks are becoming similar to those of manual workers, it has been suggested that class barriers are disappearing and either the manual worker is being assimilated into the middle class or the clerk is moving into the working class. To test this hypothesis a question on the clerk's perception of his own class position was inserted into our questionnaires. Goldthorpe and Lockwood (1963) believe that some 'convergence' is taking place between white-collar workers and working class groups, and that the two are becoming closer in attitudes and behaviour. They suggest that 'among lower white-collar workers the individualistic orientation . . . is today somewhat less pronounced than it was previously or still remains in other middle class groups'. They refute the idea that the two classes are merging, however, and suggest that while there may be 'convergence' this in no sense implies identity of attitude and interests. The clerk is still much more concerned with his personal status aspirations than the manual worker, and the latter still sees his interests primarily in terms of economic and collective goals.

One manifestation of an increased collective awareness amongst clerks is, however, to be seen in the growth of white-collar unionism. Between 1950 and 1960, for example, the Association of Supervisory Staffs, Executives and Technicians increased its membership by 109% from 11,723 to 24,551. The Clerical and Administrative Workers' Union increased by 80% from 35,130 to 59,544 and the National Union of Bank Employees increased by 78% from 29,622 to 52,787.[6] In 1963 alone the Clerical and Administrative Workers' Union recruited 6,000 members, and the largest white-collar union, N.A.L.G.O.—the Local Government Officers' union, became the sixth largest union in the country, ousting the National Union of Railwaymen from this position. This rise in membership is of course largely a product of the increase in the number of white-collar employees and this is especially true of banking, but it does show that the white collar group is now more interested in trade unionism. Why has there been this change in attitude? Wright Mills (1951) suggests that 'the acceptance or rejection of unions depends upon employees'

[6] These figures are taken from *The Times*, 18th June, 1962. Bank officials point out that the N.U.B.E. figures can give a false impression as banks were increasing their staff numbers so rapidly during this period. For example, Royal Exchange had a staff increase of 150%.

awareness of their objective problems and recognition of unions as a means for meeting them'. In the past the clerk identified more than he does today with management and management interests. As we have seen, his conditions of work were the same as those of management and he was accorded the same privileges. He was usually in close day-to-day association with his employer, he felt his job had prestige compared with manual work, and he was likely to be promoted up the managerial ladder. Because he also identified with the middle classes he adopted, and still retains to some extent, their attitudes to trade unions—that these are for the manual not the white-collar worker; that unions agitate and support strikes, a mode of behaviour alien to clerks.[7] The clerk has traditionally voted conservative and the association in his mind of trade unions and labour party politics may have been another inhibiting factor. Again, trade unions have not been part of the average clerical worker's life for very long and, where they did exist, they did not do a great deal of active recruitment. This meant that most clerks had little direct contact with trade union members although certain clerical occupations, such as local government, were exceptions here. Wright Mills believes that 'being personally in contact with trade union leaders and union members is a decisive factor in one's union attitude. In the absence of such contact and given the generally hostile atmosphere that prevails in many white-collar circles, an anti-union attitude often results'. Today most clerks have friends or associates who are union members, there is less prejudice and more pressure to join, particularly in offices where numbers of men are still employed.

Two factors may at present be both slowing down and accelerating the increase in union membership. A slowing down factor could be the large number of women and girls still entering clerical work. Women, with their short-term personal work goals, are always difficult to unionise and manual worker unions have struggled with the problem of their indifference for a long time.

[7] J. R. Dale, who interviewed 208 clerks as part of a study of clerks in industry, found that only 26 were members of a clerical union. 75% of those who were not members said they saw no need to join. 'I feel secure here,' said one, 'and I am reasonably paid. What could a union do for me?' See J. R. Dale, *The Clerk in Industry* (Liverpool University Press, 1962).

In situations such as banking or local government, where clerks are already strongly unionised, they can be persuaded to join but leadership here comes from the men. In offices where masculine stimulus is absent, union membership is probably still almost non-existent. A likely accelerating factor is the increasing number of clerks coming from manual backgrounds, with fathers who already are in trade unions themselves.

The threat of office automation could be another factor encouraging clerks to join unions and this is something we shall be examining in our analysis. Logically one would expect that exposure to such hazards as possible loss of job or reduction in status would induce clerks to join unions to protect themselves. Certainly the white-collar unions believe they have something concrete to offer their members here. The Clerical and Administrative Workers' Union in a pamphlet directed at recruiting new members makes the point that it is not opposed to automation but wants to ensure that its members get a square deal. It says:

> we shall insist on:
>
> (a) Effective joint consultation when change is contemplated.
> (b) Opportunities to acquire the new skill.
> (c) Adequate salaries appropriate to work performed.
> (d) Measures to avoid redundancy.
> (e) Compensation for loss of employment if redundancy should arise.[8]

To test the hypothesis that automation was making clerks more interested in white-collar unions we introduced a number of questions on this subject into our questionnaire and the responses will be discussed in a later chapter.

[8] The American National Federation of Federal Employees asks for even more comprehensive safeguards.

1. Thorough and definite planning ahead . . . of all departments and agencies prior to adoption of automated programmes.
2. Retraining programmes . . . for all employees *before* displaced by automation so that they are qualified for other positions.
3. Positive transfer policies—so that employees can feel assured they will be re-assigned to other jobs if they are displaced by automation.
4. Definite placement programmes for people made redundant. To place them in jobs in private industry.
5. Advance information about plans for E.D.P., which categories and how many employees will be affected and when.

These, then, are some of the social influences that have affected the clerk in the past and are affecting him today. We now go on to consider in more detail how technology in the form of mechanisation and automation has altered and is altering the nature of office work.

Mechanisation in the office

A detailed account of office machinery is hardly appropriate here and in this section it is proposed only to categorise briefly what seem to be the most important stages of mechanisation.[9] It will, of course, be appreciated that clear distinctions cannot be made between each type. The first stage is perhaps best represented by the small hand-operated machine, normally used as an ancillary or aid to the office worker. The typewriter was the first example of this to achieve general acceptance and its introduction brought about many changes in office work. One significant alteration was that the clerk's traditional art of penmanship became less important and manual dexterity replaced it for the first time as an essential office skill requirement. Today the desk calculating machine is another example of this stage of mechanisation. At its simplest it may be an adding machine which provides the clerk with a speedy and reliable means for totting up figures but can do nothing more complicated. More advanced types will multiply and divide and even print out results, although in comparison with electronic office equipment they are very small and simple and require a considerable amount of speed, accuracy and discretion on the part of the operator. These machines, although only a rudimentary form of mechanisation, can lead to considerable clerical specialisation. This can be seen in the pool of copy typists or the comptometer operator who works a comptometer on a set routine of duties all day and every day.

The second stage of mechanisation is the punched card machine. This can be based on either electrical or mechanical principles and, used with ancillary equipment such as collators and multipliers, it can perform quite a complicated series of operations. It is still basically a counting machine but it differs from the desk calculator in its size, speed, flexibility and in its method of operation. Using

[9] For further details on office mechanisation see A. A. Murdoch and J. R. Dale, *The Clerical Function* (Pitman, 1961).

punch cards it will sort, calculate and print out results. The punch card installation produces a whole new set of job specialisations. Basically there are now three distinct functions to be performed; 'pulling' the cards, punching the cards and operating the sorting, calculating and collating machines. 'Pulling' involves selecting by hand from files the cards required at any one time for punching or processing. Punching is the transposition of basic input data into a form in which it can be read by the calculator; this necessitates the use of a machine which produces holes in cards and a verifier which checks that these holes are correctly placed. Numbers are later identified by the calculator from the position of the holes in the card columns and timed impulses are transmitted to the adding mechanism and the printing unit. The third function, machine operating, is the most highly skilled of the three for it involves controlling the card processing part of the installation which, as we have seen, may be made up of a number of different machines. The nature of these jobs seems closer to factory work than traditional office work. The work of the girls is subordinate to that of the machine and the environment is often hot and very noisy.

The third phase of mechanisation—it is now usually called office automation—is the automatic digital computer. This is a logical extension of the punch card installation but it is now powered electronically and not electrically. Common to all large scale computers is the 'memory' or 'store' where information is accumulated and this is perhaps a computers' most distinguishing feature. Computers are fast, highly flexible and adjustable; they can select from alternative procedures or they can 'consult' their memory for information which has been previously fed in. In commerce or industry they have a variety of different uses, they can, for example, be used for accounts, pay-roll work, costing, personnel records, stock control, production control, sales promotion and forecasting, as well as for providing information on which management decision making can be based. Computers can be either 'special' or 'general' purpose machines, that is, they can be specially designed to perform a few fixed sequences of computing operations or programs, or they can be used for a variety of operations for each of which a new program has to be specially designed. Their use is no longer confined to the large or medium sized firm, for manufacturers are now developing midget machines which can, for example, handle the limited size payroll of a small company.[10] In fact, the criteria for the applicability of a computer to the business problems of a firm are now only two, that it must be possible to obtain a well defined procedure for the execution of the solution, and that the use of E.D.P. should result in some saving in cost or time over other methods.

Broadly, there are two approaches to the use of computers. Firstly, what Mann and Williams (1958) call the 'hardware' approach in which attention is focussed on the equipment and its ability to do a particular job. Secondly, the 'systems' approach in which the introduction of E.D.P. is seen as an opportunity for evaluating organisational objectives and procedures afresh. With the first approach, which so far appears to be that most commonly found in Britain, routine procedures such as payroll are transferred to the computer without any major organisational changes resulting. For example, many firms progress from a punch card installation to a computer and although this may result in some economies of staff and speeding up of the work, no dramatic changes occur in either the nature of the work or existing departmental organisation. Cards or tape still have to be punched, a number of machines have to be operated and a payroll department still exists. With the 'systems' or integrated data processing approach, however, a firm rethinks and rationalises its business procedures so as to use fully the potential of the computer for increasing organisational efficiency; in other words, to provide 'insight rather than numbers'. Information and decision making processes are altered and departments may be combined or eliminated as part of the streamlining process. It seems certain that in the future the chief benefits of computers will come not from automating routine procedures but from applying the systems approach and also using computers to simulate company problems and arrive at more accurate solutions. It is suggested that the managers of the future can expect to spend less time on tedious details and more time on the really fundamental parts of their

[10] In America the introduction of computers has been extremely rapid. A survey carried out by the U.S. Department of labour in 1955 showed that about 12 companies had installed machines and a further 30 were contemplating doing so. A later survey showed that in 1962 alone, 3,800 computers were installed. In this country, in 1967, around 2,000 machines are in use.

job, such as initiating action or establishing good personal relations.[11]

The implications of the computer for the clerk will be discussed more fully in subsequent chapters but some hypotheses can be put forward here for examination later on in the text. Firstly, that the basic division between manual and non-manual work is now breaking down so that when, for example, we watch a punch operator at work it is essentially a manual skill that we admire. Many office workers are now valued for their ability to handle a machine; they may have little knowledge or experience of what goes on in the office as a whole and such knowledge may even be irrelevant to the successful performance of their job. In fact, if they are located in a computer centre they may be completely isolated from the normal office world. Secondly, that the idea that computers remove the drudgery entirely from work is at present false. On the contrary, at the bottom of the clerical hierarchy in particular, the use of a computer may cause some de-skilling of work so that a routine bookkeeping job is replaced by the even more routine and boring job of punch operating.[12] However, simultaneously with this increased routinisation of work come the new and more interesting jobs of programmer and computer operator. These provide opportunities for the more intelligent clerk of both sexes and even for the manual worker, for it has been established that a capacity for logical thought, one of the most important characteristics of a programmer, is by no means the prerogative of the clerical section of the labour force.[13] In future it seems that the bulk of routine work will be done by girls with machines or by a computer. More specialised work associated with E.D.P. will be mainly taken over by men. For a while it is likely

[11] J. B. Gallagher in *Management Information Systems and the Computer* (A.M.A. 1961) criticises the fact that expensive and complex data processing equipment is often used almost exclusively for 'electronic record keeping' without any reference to the real needs of management for clear and continuing reports on the total course of a business. He believes this type of operation is self-defeating.

[12] This kind of routine work will go as computer technology advances.

[13] A great deal of research is being carried out by computer manufacturers and others to establish the best methods for identifying potential programmers. Some manufacturers favour a battery of intelligence tests; others tests based on job simulation.

that staff for these jobs will be recruited from existing male clerical ranks, but as these diminish and technical colleges start offering school leavers full time courses in computer techniques, the computer specialists of the future may have and need no general office experience at all.

Some problems of change

We have now considered briefly the changing role of the clerk in industry and the nature of the mechanisation that is affecting his job. Before concluding this chapter it seems important to make some general statements concerning organisational change and to suggest some lines of approach to the problem of ascertaining how the kind of social change arising out of the use of computers may affect a firm and the individuals working in it.

Most organisations by the nature of their structure are geared to stability or slow change. Accelerated change of the kind that can result from a take-over by another company, a re-organisation of production processes, or the introduction of a computer into a firm's offices, will usually prove disruptive and disturbing. This is because people in stable organisations are adjusted to and expect a steady continuation of existing arrangements and also because organisations unaccustomed to change frequently find they have become too rigidly structured to assimilate it without stress (Mann and Neff 1961, Sofer 1961). It has been suggested that organisations will vary in their capacity to accept change depending upon their flexibility, the nature of their human resources and their understanding of the consequences of change. Factors relevant to successful adaptation to new techniques would seem to be the climate of relationships that has developed over the years between management and employees, the firm's previous experience of change, the amount and kind of managerial talent and the recognition by management that technical and organisational change must proceed simultaneously, the kind of skills to be found in the work force and the age, sex, expectations and objectives of staff likely to be affected. These will be considered in the analysis of our two firms. An additional factor which may contribute significantly to the ease with which an organisation can change to E.D.P. is the degree of pre-change mechanisation already in existence although, against this, is the possibility that some experience of mechanisation may lead to an under-estimation of computer problems. Our firms provide an interesting comparison here in that one was changing to E.D.P. from a punch card installation, the other was changing from largely manual methods of data processing.

One of the most interesting and least investigated aspects of E.D.P. is the effect it has on relationships within a firm. Any organisation is a social system built up of a network of relationships and change in one part of the organisation or in a particular set of relationships will have repercussions in other parts of the organisation and on other sets of relationships. Because rapid change means upheaval, uncertainty and the overthrow of custom and tradition, it tends to make people more aware of the nature of their relationships with colleagues, managers and subordinates. It therefore provides the sociologist with a particularly good opportunity for studying group behaviour and for identifying and examining those relationships which are essential to the successful running of the organisation and the attainment of organisational goals. Advanced technological change can result in changes in the power structure of a firm and in a weakening in those relationships which before the change were considered to be crucial to effective operation.[14] There may also be a variation in the kinds of relationship that are needed for efficiency. For example, during the change period technical knowledge may become more important for supervisors than human relations skills.

The individual and his reactions to change must also be considered and the office worker has particular problems here for as a rule he has had less experience of change than the shop floor worker and may therefore find it more difficult to accept. How an individual receives change will depend on his previous experiences, the social and work groups he belongs to, his personality, his age, his plans and aspirations, and what consequences he believes the change will have for him personally. It may be hypothesised that young people will welcome change more than older age groups for it can provide them with a challenge and stimulus and with opportunities for self advancement. If pressures become unpleasant they can move to another job in a different firm

[14] The computer gives power to new and often unsophisticated groups. A strike of punch girls, for example could paralyse a highly computerised firm.

without much difficulty. For long service employees it is more of a threat. They may be unwilling to adopt new and unfamiliar methods of work, the need to learn new techniques may be unpleasant and stressful and, as a rule, there is no way out of the situation for their chances of finding new jobs are slight. Problems associated with retraining long-service staff and helping them adjust to new conditions are therefore important areas for research.

All employees, however, are likely to have some anxieties when they first hear that their firm is installing a computer and the principal cause of these is the association of E.D.P. with redundancy[15]. Computer manufacturers stress the possibility of considerable staff saving as part of their sales campaigns and most firms hope that their new machines will reduce labour costs even though they are confident that they can absorb displaced staff elsewhere in the organisation. We were particularly interested in finding out whether any redundancy occurred in our two firms and also the amount of job displacement that takes place in departments whose work is taken over by the computer. Even when redundancy is avoided it seems likely that the employment difficulties of school leavers may be accentuated if firms, once they get a computer, make drastic cuts in the numbers of school leavers they engage each year for office jobs.[16] This is another point we shall need to examine. Finally, any study of social change, besides considering its effect on the organisation, the individual and the group, must also look at the kinds of strategies employed by management to

[15] The 1955 American Congressional Sub-Committee which studied the impact of automation on long-run employment and investment levels reported '. . . no one dare overlook or deny the fact that many individuals will suffer mental and physical hardships as the adjustments go forward.'

[16] *U.S. News and World Report*, May 13th 1963, states '. . . the computer is a growing source of worry to many high school guidance counsellors and to the U.S. Department of Labour . . . Official figures indicate that few new clerical jobs, apart from stenographers, secretaries and typists, opened up in the two years from February, 1961, to February, 1963. The increase was less than 2% in this two year period, when business and employment generally were rising from recession levels. A study of jobs in the Insurance industry indicates that employment increased by 200,000 or 23% in the five years ended with 1957, when computers were just beginning their invasion. In the next five years, employment in this field increased by only 80,000 or less than 8%.'

implement the change and secure its acceptance. Mann and Neff (1961) suggest that 'the fundamental task of any group responsible for introducing change is to devise a plan that will bring together organisational objectives and individual goals.' It will be important to ascertain whether the management of our firms were able at an early stage to identify staff needs and goals and to make plans to meet them; we will also need to examine how they communicated information about E.D.P. to staff, the extent to which they felt consultation to be necessary and their success or failure in 'selling' the idea of a computer to their employees. Although the most appropriate strategies to adopt will differ from firm to firm according to the kinds of attitudes and values that prevail, it may be possible from our data and that of other studies in the same area to suggest some general principles which will help firms to reduce the amount of stress arising out of changing technological systems.

PART II

Our case study firms

CHAPTER 3

The Royal Exchange Bank—The development of the new computer centre

Historical background

At the beginning of this century much of the work of a branch bank consisted of account keeping. There were few mechanical aids and all ledgers, passbooks and statements had to be written up by hand. Bank clerks were exclusively male, and their principal skills were an ability to handle figures and to write legibly. Starting salaries were £30 a year and prospects were limited. Today this situation has greatly altered. Almost half a bank's clerical staff are women and promotion opportunities, for men at least, are excellent. Staffing advertisements claim that one in two male entrants to banking are likely to become managers or executives; in 1964 a junior executive in his late twenties or early thirties could expect a minimum salary of around £1,350, while young managers would receive considerably more.

Banking graduated from hand kept ledgers to mechanical accounting machines in the 1920s. From the thirties onwards mechanisation continued to spread and develop, stimulated after the war by an acute shortage of suitable staff, together with a continuing increase in business. By the 1950s American banking experience was suggesting that the use of electronic data processing techniques might provide a solution for these staffing and expansion problems, and it was during this period that many British banks ordered their first computers. Besides reducing the demand for staff it was believed that the computer would eliminate much of routine clerical work, thus providing speedier promotion for the male entrant. This last point was of great importance at a time when competition for school leavers was making it exceedingly difficult for banks to obtain recruits of the right quality. Earlier mechanisation had already reduced the routine work of the male clerk; the computer was seen as continuing this process.

Royal Exchange's decision to install a computer

In 1955 the Committee of London Clearing Bankers set up a special sub-committee to consider electronic aids in banking. The attention of this new committee was directed primarily at the development of a common machine language acceptable to all banks so that cheque reading could be done mechanically. It was appreciated that if data could be automatically taken off a cheque, this would revolutionise sorting procedures at clearing houses, and also enable individual banks to fully automate their ledger posting procedures. It was at this time that the manager of the Royal Exchange Organisation and Methods Department became interested in the potential of computers for banking and encouraged his own management to consider the idea of installing one of these machines.

Royal Exchange senior management decided to pursue this idea and towards the end of 1958 detailed work specifications were drawn up and made known to all computer manufacturers. Feasibility studies were made jointly by the bank and six interested manufacturers, and using these as a guide a suitable model of computer was provisionally selected.

The next step was to make a trial accounts run on a machine similar to the one chosen, and a small London branch was used to provide data for this experiment. A computer program was written and the accounts of this branch were processed electronically at the computer manufacturer's service bureau. This test confirmed that bank records could be handled by a computer. In January 1960 a demonstration of the new system was given to the press and to Royal Exchange directors and general management. After this a definite order was placed and the new machine was scheduled for delivery early in 1961.[1]

It was expected that with a computer and an operating staff of five per shift, accounts could be up-dated at the rate of 6,000 an

[1] The data processing installation ordered was a basic computer, plus two punched tape readers; two tape punches, one fast and one slow; a magnetic tape console, and four magnetic tape units. The computer used a binary number system. Its main store had an 8,960 word capacity. Data input was read at 300 characters a second, data output was punched at 300 characters a second on the fast punch and 60 characters a second on the slow. Data punched on tape was converted into print at the rate of 10 characters a second.

hour, and that the computer would be capable of handling the accounts of approximately 30,000 customers. The existing system required the use of 80 business machines and a staff of 100 to keep the records of this number of accounts.

The new machine's book-keeping programme would take care of eleven main accounting procedures and two subsidiary ones. Among the banking operations to be performed were the entry of normal credits and debits, the alteration of basic account information, the addition of new accounts, the deletion of closed accounts, the accumulation of statistics for management and the daily printing of the current balance of every account. Statements would be printed out when required.

Royal Exchange decided to buy the computer outright rather than to rent it. As a matter of policy the bank did not like renting equipment, and management believed that income tax allowances would help to keep down the purchase cost. This first computer installation was seen as the initial phase of a long term change process which would culminate in the general use of automatic accountancy procedures throughout the bank's organisation.

Preparing for the computer's arrival

Once the new computer had been ordered the Royal Exchange O. & M. staff were faced with a great deal of work. Systems analysis, programming, and planning the conversion to a new system all needed to be completed before the computer arrived. In any firm, systems analysis as a preliminary to electronic data processing will normally involve a survey of existing operations to determine which are suitable for transfer to the computer, together with the creation of a new system designed to utilise the potential of the machine. This is a lengthy process as it necessitates a complete understanding of the functioning of every detail of the old system, and the selection of a new system which will not only replace, but if the computer is to justify itself, improve on the old. The complexity of the task will vary considerably with the nature and extent of the work to be processed electronically and especially with the degree of re-organisation involved. In some cases a computer may be used to take over just one group of routine processes, all part of the work of a single department. At the other extreme there will be centralisation, the re-allocation of functions between departments, and the re-organisation of large areas of work. Examples of major re-organisation seem to be more characteristic, so far, of American than of British experience but this is largely due to the length of time American industry has been using computers. In this country current practice is still for firms to use computers for speeding up and streamlining routine functions. Here the task will be made easier if the functions to be taken over are already partially or completely mechanised through the use of a punched card installation.

It goes without saying that this phase of the preparation period is of crucial importance since, without a good system, the new computer is not going to pay for itself. The staff of the computer manufacturer is normally available to give assistance, but, although they can give much useful information and advice, it is up to the firm's own technical experts both to master the details of the new system, and to make sure that it adequately fulfils the needs of the organisation as a whole.

The scope of the re-organisation was fairly extensive in Royal Exchange. Although the computer would take over only routine clerical operations, this involved the centralisation of the accounts of a number of branches and the transfer of these accounts to the computer centre. Also, the transition from traditional to automated banking necessitated taking in one step the jump from accounting machines and manual clerical procedures to a computer without the intervention of a stage involving punched card equipment.

The nature of banking procedure set its own problems. One of the main requirements of the new system was to maintain the accuracy of the old by means of a series of checks. It was also essential that the data processing operation should keep to the same daily time-table as before. This emphasis by the bank on a system giving a high degree of accuracy made for difficulties later, in that it necessitated a more complex, and also a slower, procedure than would otherwise have been required. There was some difference of opinion between the bank and the staff of the computer manufacturer as to the need for such eleborate checking devices but the bank was not prepared to give way on this point. Whereas small accountancy errors might not be of great importance in most commercial houses, the reputation of a bank depends on the confidence its customers have in its ability to handle their affairs with scrupulous exactness. Because of this, Royal Exchange was not prepared to risk any errors in customer accounts.

Although the bank was anxious to retain as many aspects of the old system as possible, there were some procedures which had to be abandoned. The practice of issuing statements on demand could not be continued with automated banking. Customers were to be asked to give the bank twenty-four hours notice so that statements could be prepared in the computer centre. A further difficulty arose from the association, in the minds of both customers and staff, of the computer with the bank's decision to abandon the use of full narrative statements.[2] Instead of setting out in detail the

[2] Below are examples of narrative and non-narrative statements. Non-narrative statements have cheque numbers instead of names.

DATE	PARTICULARS	DEBIT	CREDIT	BALANCE
1965	Forward			48.10. 5
7 Jul	Harper	6.16.0		41.14. 5 cr
11 Jul	Bimpson	7.18.3		33.16. 2 cr
12 Jul	S.O. Utd. Equit. Mortge	5.13.8		28. 2. 6 cr
19 Jul	Div. Robt. Transport		13.10.0	
	P.M.G.	22. 5.8		
	Self	15. 0.0		4. 6.10 cr

DATE		DEBIT		CREDIT	BALANCE
31.6.65	FWD				48.10. 5
7.7.65	233	6.16.0			41.14. 5
11.7.65	235	7.18.3			33.16. 2
12.7.65	S.O.	5.13.8			28. 2. 6
19.7.65			DIV	13.10.0	
	234	22. 5.8			
	236	15. 0.0			4. 6.10
24.7.65	238	5.10.0			
	CQB	3.4			1. 6. 6 dr
25.7.65	239	14.15.0			16. 1. 6 dr
29.7.65			C.T.	100.14.3	
	237	13. 4.9			
	240	5. 3.4			
	241	10. 0.0			56. 4. 8

names of payers or payees, cheque numbers alone were now provided and a customer would only know if a particular cheque paid out by him had been cashed, or if he had received an outstanding debt, by keeping a personal record of all the cheques he sent out and received. In fact, this change from full to limited narrative had nothing to do with the computer but it did cause the new automated system to be regarded more unfavourably than it should have been.

The next phase is that of programming. This was an extensive procedure at Royal Exchange as, at the time of the first installation, computers were still in their infancy in this country. The computer chosen, a commercial version of one originally designed for scientific purposes, was not auto-coded and all working instructions as well as basic data had to be fed into it. (This would still be true today as computers with built in bank programmes are not yet available.) Programming is difficult, not so much because it involves the use of complicated techniques, but because the logical step-by-step analysis of a particular procedure necessary to formulate instructions for a computer requires a most exhaustive preliminary knowledge of that procedure. Computer manufacturers therefore frequently advise their customers to recruit programmers from amongst their own staff as these will already have this preliminary knowledge. Royal Exchange decided on this policy and all the programmers were ex-bank clerks who volunteered for the job.

Although a great deal of assistance was provided by the computer manufacturer, this was a difficult period for the bank, involving as it did, not only the initial writing of programs, but constant testing and revision. It seems likely that the inexperience of the programmers contributed to the difficulties of this phase. Many of the computer staff spoke of it as a period of anxiety and strain, although it must also be admitted that for some of them this was compensated for by the challenge and excitement of the job. The process of re-writing programs continued until well after the computer arrived for it became increasingly clear to those carrying out the task that although their first programs were adequate, further work could make them both better and faster.

The next and final process is that of planning the actual conversion to the new system and this has to be completed in the pre-installation period if there are not to be serious and expensive delays later. The extent of the work involved will vary from firm to firm, depending on the nature and scope of the computer installation, but it will normally include deciding the sequence in which work is to be transferred, organising the preparation of basic input data, designing or amending documents to fit the new techniques of information, running down the labour force and making arrangements for the re-allocation and re-training of displaced staff, and recruiting and training any new specialist staff which are required. If there is a parallel running period with old and new systems operating side by side, firms may need to take on temporary staff to help cope with the additional work load this involves.

Royal Exchange made its plans for conversion well in advance of the arrival of the computer. These provided for a period of parallel running followed by the transfer of work to the machine in a series of stages. In practice the time-tabling of events as set out on paper proved to be wildly optimistic, and schedules had to be abandoned. The reasons for this will be discussed later, but it is clear from the very great disparity between the preparatory time estimates and the actual dates on which branch work was transferred to the computer, that the bank was quite unaware, at this stage, of the technical problems that it was about to face.

No advance plans were made for the re-allocation of staff in automated branches for, although some redundancy was anticipated, the normal process of wastage was expected to take care of it. Indeed, a greater source of anxiety was how to manage the extra work involved in the period of parallel running. Some plans had to be made for staffing the computer centre with operators for the machine and its auxiliary equipment, but the appointment or transfer of these operators, and their training, was not undertaken until after the computer had arrived, the programmers acting as computer operators during the intervening period.

A great deal of the preparatory work took place in the branches, and not the least of this was the process of cheque personalisaticn. This involved the printing of the customer's name and account number on each of his cheques. It had to be done for all customers attached to any branch transferred to the computer as the machine used the account number for identifying particular accounts. Cheque personalisation was carried out under the supervision of the bank's research staff, but extra clerical staff were drafted to the branches to provide additional help. The preparation of basic data for the computer had also to be done in the branches, and this too involved a great deal of work. The computer team first went through every branch account, listing all data required for the automatic processing of these accounts. This was then read into the computer and two files set up on magnetic tape, a balance file and a history file. The balance file contained the customer's account number, his name, current balance, account limits—that is the amount of overdraft the bank would accept—and various accumulated statistics for the use of management. On the history file were again recorded his account number and name together with the periods at which his statements of account were to be produced. This file also had a note of the balance and date on his last statement, and the date, nature and amount of each transaction that had occurred since this statement was supplied. Once the new system was in operation these two magnetic tape files would be up-dated once a day.

The acquisition of all this information was a time consuming job and the computer centre staff had to spend long periods in each branch. In fact, only one branch was completed when the computer arrived. Although programs were written as completely as possible, inevitably some account information became out of date and this meant that the computer team had to return to the branches on a number of occasions in order to get additional data. This tended to cause strained relationships between branch and computer staff, particularly in the first branch scheduled for transfer where the manager, although prepared to accept the new system, was not at all enthusiastic about it.

One further aspect of the planning stages must be indicated. Apart from the computer itself, any computer installation also involves ancillary operations of which the most important are the day-to-day preparation and feeding in of input data and the disposing of output from the system. These ancillary operations will involve the choice of equipment such as punch machines, the planning of procedures for each operation and the selection and training of personnel. Royal Exchange managers were presented with a number of problems here. Not only were they quite unfamiliar with punching equipment, but the input to their computer was in the form of paper-tape. While this has its advantages, the development of equipment in this field was, at the time of installation, not so advanced as for punch cards and many of the bank's early punch machines proved to be unsatisfactory. Also the bank had no experienced punch operators. This meant that girls who had previously been ledger posting machine operators or on general clerical duties would have to be trained to prepare input data for the computer.

The installation period at the computer centre

The computer arrived in the Centre at the end of March 1961, several months after it was expected, and from then until July it was in the hands of the engineers for testing purposes. In July parallel running commenced—that is the processing of accounts by the computer simultaneously with existing procedures. This had two functions: the first to check that the computer program was correct and adequate, and could cater for any circumstances likely to arise; the second, to enable branch personnel to establish that the computer could supply them with all the information they required.

The period of parallel running caused some friction in the first branch to be transferred because of the considerable amount of additional work it caused. Data had to be specially produced for the computer while staff were still required to continue with their normal jobs. 'Getting away early', a concession much valued by the branch staff, became very difficult and this led to many comments on the disadvantages of the new system and unfavourable comparisons with as one clerk put it 'the old fashioned way of doing things'.

In August the bank announced in *The Guardian* that 'the period of operating the computer in parallel with normal branch work is over and both branch and customers are now dependent on the computer for their records'. This statement was somewhat premature as it co-incided with a major breakdown at the computer centre which put the computer out of action for a number of days and had a most damaging effect on the programmers' morale.

Progress now began continually to be held back by machine breakdowns. These affected both the punch machines at the branch, the computer itself and the ancillary equipment attached to the computer. The completion of data processing in the evening was constantly delayed and the programming staff began to work extremely long hours. They became both harassed and anxious. Not only were they struggling with equipment which was technically unreliable, but were also suffering from fatigue and overwork.

However, in October, despite the difficulties that were still being encountered, running in parallel was completely terminated and the Centre became entirely responsible for processing the work of the first branch. Each afternoon tapes containing details of the day's transactions were sent over to the Centre after the closure of branch business. This meant that data processing in the computer centre could not commence until around 4.30 p.m. Unhappily for the computer centre staff the breakdowns continued, yet, because of the nature of the bank business, one day's work had to be finished before the next was started, and it was not unusual for programmers to have to stay until the early hours of the following morning in order to complete a particular day's work. Weekend work also became a necessity in order to catch up with scheduled programming and to find opportunity for re-testing programs. In these early days there was no formal shift work system in operation and no staff, other than the programmers, to operate the equipment.

A representative of the computer manufacturing company, when asked the reasons for this continued overtime working, suggested that many of the holdups were due to internal organisational defects. She felt that work at the Centre could be speeded up considerably; for example, bad tapes from the branches ought not to be re-done at the Centre, but sent back to the branch of origin. She also maintained that programmers should not operate the computer at all, as trained operators were more efficient. The latter worked to a set routine, whereas programmers, as she put it, 'mess about and investigate imaginary faults'. However, she did agree that the Centre had had a lot of machine trouble and that some of this was due to faulty parts being sent from the manufacturer's factory.

By November 1961 the computer was becoming more reliable. Breakdowns were no longer so numerous or so serious, although they still caused some delay and proved a source of irritation to the Centre staff. It was now decided to make plans for the work of the second branch, a much smaller one, to be transferred. This transfer was completed by Christmas and further plans were then made to transfer the work of the third branch by the middle of

February 1962. There was now a number of girl assistants in the computer centre and with their help the computer staff began recording the basic input for this branch. This time, instead of computer staff visiting the branch, customer data was sent over to the Centre and the information for the balance and history files was punched into paper tape there.

Simultaneously with these developments the staff of the Organisation and Methods Department were evaluating the ancillary equipment associated with the computer. They were not at all happy with the branch punch machines, which were very liable to make errors as a result of mechanical defects, and suggested that the bank should not confine itself to one make, but experiment with the models of different manufacturers. Eventually, after many trials, a punch machine was introduced that incorporated a large number of checking procedures. This machine provided a parity check, an account number check and 'answer back' circuitry, a device for checking that the punch is reproducing the symbols indicated by the girls on the keyboard. These machines proved most successful and were found to be more reliable than the machines installed at the first two branches.

Plans had earlier been made to obtain a line-printer, a machine which, when connected to the computer, would print out balance listings automatically. This would speed up data processing considerably as the existing system involved printing out data from tape on separate machines. It was anticipated that the line printer would arrive in January 1962 and the Centre staff, in addition to refining existing programs, commenced working out a program for this. They also had to re-write the computer's history file as the original program was written on magnetic tape in 32 word sections, whereas the line printer required sections containing only 16 words.

The computer was now running reasonably satisfactorily and, providing work was not held up by breakdowns, the two branches on the computer could be completed by 8 p.m. The girl assistants were not permitted to work beyond this time, any unfinished work being completed by the supervisor and the duty programmer. Three branches could not be completed by 8 p.m. however, and it was envisaged that when the third branch was transferred the finishing time of junior staff would have to be changed from 8 p.m. to 10 p.m.

Unfortunately, plans to commence processing data for the third branch by the middle of February had to be postponed as the manufacturers had not been able to deliver the branch punch machines by the agreed date and the staff could not produce the tape required by the computer. However, this branch was successfully transferred in March and it now became necessary for all the Centre staff to work until 10 p.m. But work was in hand to streamline computer operations and it was believed that a new series of procedures evolved by the programming staff and designed to incorporate the line printer would eventually cut down the time required for processing each branch.

By October, 1962, therefore, the computer was processing the work of three branches and, in addition to this, keeping records of all travellers' cheques issued by the bank. The computer staff now began to get ready to take over the work of the fourth branch, anticipating that the line printer would be in operation in a few weeks time. However, once again, an unanticipated delay occurred and plans had to be abandoned. In actuality the work of the fourth branch was not transferred until April, 1964, eighteen months later, as the line printer proved to be both mechanically and electronically unreliable and it took many months to rectify its defects. Two and a half years had now elapsed since the work of the first branch had been transferred.

It had always been envisaged that eventually the programmers would cease to concern themselves with the actual day-to-day operation of the computer and would devote their time to programming and development work for future needs. By 1963 this was becoming possible. They were now able to give much more of their attention to experimental work, leaving the routine day-to-day data processing in the hands of girls who had been trained as operators. With the exception of the line printer, the new streamlining procedures were now fully operative and the Royal Exchange Organisation and Methods Department began turning its attention to the development of automatic banking in the London area. Logically London was a more appropriate centre for data processing, once the problems of data transmission from branch to computer centre had been solved, as 70% of the bank's cheques passed through the London Clearing House.[3]

[3] Data was taken from the branches to the computer centre by messenger. However, the bank was experimenting with methods of automatic data transmission.

Because of this shift of interest to London and the purchase of a new computer which was to be sited there, a decision was taken to abandon the original plans for the fifth and last branch, and to transfer only a very limited amount of its work. This unexpected development presented the university research team with a serious problem, for they had expected that this branch would be fully computerised by July 1964, and that post-change interviews would be carried out there in January 1965.

Staff and staff relationships

In the preceding sections we have touched briefly on the computer centre staff and their problems. Let us now consider this group in more detail, seeing the kind of people they were and their attitudes to their work.

PROGRAMMERS

Once the computer had been chosen and ordered the Organisation and Methods Manager had to build up a team of systems analysts and programmers which would plan and carry out the changeover from the old system to the new. He decided not to split up systems analysis and programming between different individuals but to make the new programmers responsible for the two procedures. As ex-bank clerks they would be competent to complete the systems analysis, for their knowledge of traditional banking procedures would help them with the necessary break down of existing work methods and the design of new ones.

The Organisation and Methods Manager already had a staff of two, for the first appointment had been made as early as 1956 in anticipation of a computer being ordered. Later, when the bank started thinking seriously about feasibility studies, another man was added to this small group although he left after a short while and the first man completed the experimental program alone. Other programmers were selected well in advance of the computer's arrival, one at the end of 1959 and three more during 1960[4]. Unfortunately the original member of the team, by now well experienced also decided to leave the bank in the summer of

[4] This team was concerned with preparing for the first computer. Other programmers were recruited later in connection with the bank's future plans.

1960. This was a great blow to the O. & M. Manager and to the programmers who were now left, as one of them put it, 'to learn from our own mistakes.'

All of these programming appointments were made from local staff already employed by Royal Exchange. Staff selection was not based on aptitude tests, as there were few of these in existence, but on recommendations from senior management and requests for consideration from interested bank clerks. At this time little was known about the qualities of a good programmer and the bank believed, probably rightly, that a most important consideration must be the ability of the team to work harmoniously together.

All but one of the new group were men and their ages ranged from 26 to 37. All had had a grammar school education and several had been in the Organisation and Methods Department as mechanisation officers before being transferred to the computer staff. The one woman was an interesting exception for she had been a supervisor of machinists before being chosen to take part in the programming. Initially it was intended that once the installation of the computer was completed, she would be the supervisor in charge of female staff at the computer centre. However, after the arrival of the computer, this scheme was dropped and she remained on programming, another supervisor being appointed.

Members of the new team were attracted to programming for different reasons. The supervisor saw her job vanishing with the advent of electronic data processing. One of the men thought he would enjoy the variety of work, another liked the entirely new approach to banking problems, and a third became interested in computers when studying economics for a professional examination. The team's formal training in programming was short. Some months before the computer arrived they attended a two weeks course run by the computer manufacturer and they also spent two days a week in the manufacturer's computer centre testing out the newly written programs. The rest of their training was acquired on the job, learning from each other and by trial and error.

The programmers, although they had so recently been bank clerks, showed behaviour patterns and attitudes that appeared to differ from those normally to be expected in bank personnel. It seemed that once they left the environment of the branch, with its pressures on staff to conform to a particular bank image, they felt that their role had changed from clerk to that of 'boffin'. Dress became less formal—it was common for men to work in their shirt sleeves—and there was a free and easy social atmosphere. Loyalty ties to the bank seemed to weaken as the team realised that they were acquiring knowledge and skill with an outside market value, and that personal career aspirations were no longer dependent on success in the banking world. One programmer, who had been with the bank since he left school, said that he now felt competent to programme most administrative procedures and that if a suitable job cropped up in industry he would take it. He was only staying with the bank to gain experience as he felt the small amount of money the bank paid programmers, over and above the normal salary scale, did not make the job worth while. This programmer believed that many of his colleagues shared his views on the unsatisfactory financial rewards of the job and the fact that the two first programmers left so soon after their appointment lent support to his opinion. At this stage the bank did not appear to appreciate the importance of its programmers, nor did it anticipate the intense competition for this kind of skill that was to develop in industry generally. It apparently assumed that programmers would follow the traditional behaviour pattern of the male bank clerk, that of staying with one bank for life, and it was unaware of the necessity for making programming and computer work a really attractive job.

Programming teams would seem to be fruitful areas for inter-group conflict as individuals jockey for position and power in an, as yet, unstructured situation. However, there was no evidence that this kind of conflict occurred in the bank team who seemed able to work most amicably and co-operatively together. But, although there was no conflict within the team, during the early days, with their crop of technical problems and the need for extensive overtime work, the programmers appeared at times to be unhappy with certain aspects of their work situation. For example, when technical problems were at their height some of the programming staff felt that top management was not sufficiently aware of the numerous breakdowns that were affecting the computer and of the consequent difficulties the computer centre staff were having to face. They believed, perhaps incorrectly, that management lacked understanding of their problems. They had the entire responsibility for dealing with computer failure and making up any resultant backlog of work. Although they were making superhuman efforts, these were not recognised by top management which, intheir opinion, dissociated itself from the difficulties of running the Centre.

Pressure from above for faster results and the belief that there was a lack of understanding of the problems they were facing greatly increased the programmers' feelings of stress and anxiety. They were particularly frustrated at their inability to communicate their problems to anyone in authority. The O. & M. Manager had, at this time, to devote a great deal of his attention to future needs and preparing plans for computer development. This inevitably restricted the amount of time that he, personally, was able to spend in the computer centre. One member of the computer team suggested to us that a computer centre manager needed to be appointed at once in order to deal with problems as and when they arose, and to make top management understand that there were problems.

This kind of stress situation seems fairly common when a team of innovators are grappling with a new problem of technical change. In chapter 6 we shall see how it was repeated almost exactly in our second firm and there are numerous examples in American research reports. Mann and Williams (1960) describe how in one American company introducing a computer, pressure mounted until fatigue and stress reached a climax for key personnel. One executive had a heart attack which brought to everyone's attention the tempo and level of tension experienced by those directly involved in the change.

GIRL ASSISTANTS

The girl assistants were employed to operate the computer and the ancillary equipment—including the teleprinter, the machines which printed out balance information and statements, and those punch machines that were kept at the Centre. They were all newly recruited school leavers, nearly all had had a grammar school education, and they arrived in the computer centre in the second half of 1961.

These girls came under the control of a supervisor transferred from the bank's training section who had very considerable experience both of banking and of training needs and procedures for conventional ledger posting machines. This supervisor was helped by the two young assistant supervisors, each with two years experience of banking, one in a branch and the other in the bank's clearing house. It was decided that, at first, these two girls would help operate the computer in addition to their supervisory duties.

Initially the girl assistants presented a number of problems. First, and most serious, was that of recruitment and training. The bank, although knowledgeable about the training of conventional machinists had, as yet, little or no knowledge about the requirements of its computer installation and the kind of personnel best suited to this kind of work. Traditionally the bank had always recruited girls with a General Certificate of Education, but there was some doubt as to whether this level of intelligence was required by the relatively simple machine minding jobs in the computer centre. A second problem was the necessity for shift work in the Centre, a situation entirely new to banking. These young girls would have to work shifts as data could not be processed on the computer until after the branches had closed at 3 p.m.

By October 1961 the computer centre had been allocated fifteen girls, all of them straight from school. This was a particularly difficult period for the supervisor in charge. She not only had to learn about and help operate the computer herself, but also train this large number of girls in these very new jobs. The computer and ancillary machines were still proving technically faulty and the supervisor complained that every time she tried to instruct the girls there was a breakdown. This made life frustrating for her and boring for the new trainees, as for long periods of time they had little or nothing to do. In addition to this difficulty there was not sufficient space in the Centre to contain such a large group and the machines were too few in number to provide adequate training facilities. In view of this the training plan was revised and three girls at a time were sent to the Centre for periods of two weeks. The girls who were not on this two weeks course were loaned to various branches until such time as they were needed by the Centre.

When all the girls had been through the two weeks training period they were allocated so many days a week to attend the Centre, the rest of the time being spent at a branch. A two shift procedure was introduced with some girls working a day shift from

9 a.m. until 5 p.m. and others working an evening shift from 3 p.m. until 8 p.m., leaving earlier if the work was completed. The introduction of a rota ensured that any girl had only to work the late shift one day in three. The assistant supervisors participated in this rota but the supervisor adopted the same system as the programmers. That is, she worked from 9 a.m. in the morning until 5 p.m. at night, and also stayed late several times a week to help with the data processing. One problem caused by the overlapping shift system was that between 3 p.m. and 5 p.m. there were, in addition to the supervisor, nine girls in the Centre—six trainees, two assistant supervisors and one permanent staff—and this was a time of day when there was very little for these girls to do.

By November the organisation had been modified further. Six girls were now permanently attached to the Centre and these worked there full time. Each day four of these girls were on the day shift, helping to produce statements and to prepare basic in-put data, and assisting with the computer and balance lists. The supervisor felt that this was a better system than the original idea of keeping a large number of girls attached to the Centre, but lending them to the branches when they were not needed.

By January 1962, this system was working quite smoothly. In addition to the supervisors, who had now become experienced operators, a third girl had been taught to operate the computer. The first two branches were now on the machine, the data for the first arriving at 3.30 p.m. and being completed by 4.30 p.m. There was then often a waiting period until tape from the second and larger branch arrived. The Centre was, as a rule, able to complete the work of both branches by about 8 p.m. and, in fact, the staff were doing more work than was strictly necessary for they identified errors in tape sent over from the branches, made corrections and reproduced the faulty tape. They continued doing this for some time although it had originally been agreed that when a branch sent over incorrect tape, the Centre would place this on one side and return it the next morning to the branch concerned for correcting.

During this period the training of the girls continued and all the girls were soon able to perform the essential tasks, including operating the computer. In March the third branch was transferred on to the machine and the girls were asked to work until 10 p.m. The bank, anxious in case this proved tiring, proposed that as soon as three new girls could be acquired they would only work late one night in four. It was also anticipated that once the work of the fourth branch was taken over, a Saturday shift from 8.30 a.m. to 4 p.m. would be introduced. This would mean that each girl would have to work every third Saturday (Saturday morning work is of course normal for bank clerks and so this was not an innovation). On the whole, the girls did not object to working shifts. They were given time off to compensate and were also paid an inconvenience bonus. Some liked the extra time they had during the week as this enabled them to do their shopping and visit the hairdressers.

It is interesting to note that eight of the nine girls in the computer centre in 1964 had had a grammar school education. The bank apparently believed that grammar school girls were as well suited to the new computer jobs as to the traditional branch ones.

Unfortunately we have no systematic information on the attitudes of the girl assistants to their work as their supervisor proved unwilling to allow them to be interviewed.

Relations between the computer centre and branches

During this period the computer centre staff had the extremely delicate task of maintaining good relationships with the staff of the branches whose work was being transferred. This was no easy task as the branches, and in particular the first branch taken over, believed initially they had little to gain from being mechanised and that they might have a considerable amount to lose. In any event they were certain the process of transfer would cause them much disorganisation and additional work. The attitudes of branch staff to the way the change was handled will be described in detail in chapter 5 but one or two points can be usefully made here.

The technical difficulties encountered in the early days certainly affected the work of the first branch to go on the computer and led to a deterioration in relationships between branch staff and computer personnel. Parallel running, involving as it did the operation of unfamiliar machines, caused a considerable increase in work and branch staff found that they frequently had to stay late in the evening to cope with data preparation.[5] Because of this, in-put tape was often late arriving at the computer centre and the computer operators tended to feel that this lateness was due to lack of co-operation on the part of the branch. Staff at the branch, for their part, complained about the necessity for overtime working, and criticised the Centre for being unable to get customers' statements to the branch by 12 a.m., the time agreed on as a deadline for this procedure.

Another major problem in the first months after transfer was the production of inaccurate in-put data by inexperienced machinists who were very prone to make mistakes with the new machines. Errors in amount totals were not very serious as these always showed up in the final computer balance but a greater difficulty was the likelihood of girls transposing a correct amount to a wrong account number. As we have seen, for quite a while the computer centre staff checked and corrected faulty tape sent over by the branch but they finally came to believe that the number of errors would only diminish when the branch was made responsible for checking and correcting its own tapes. In fact, as time went on, the branch machinists did become more skilled at the new work and errors decreased. Gradually branch personnel became reconciled to the new system and although a number of senior staff were slow

[5] Firms do not always appreciate that more not less staff are required during the period of parallel running when old and new systems are operating side by side. The following cautionary tale appeared in the *Observer* in January, 1965.
'Gloucester today is an unhappy city. Its two chief officials—the Town Clerk and City Treasurer—have retired prematurely, and another official in the treasurer's department is appealing against being asked to find a new job.

Besides these human tragedies arrears in mortgage repayment of more than £28,000 seem slight . . . The accountants' "Report regarding certain matters concerning the City Treasury" shows that 188 people . . . have been or are more than 3 months in arrears with their repayments. It also reveals that half last April's rates demands were not sent out and that no ledgers, controls or cash books were opened for the 1963-64 accounts.

The cause of all the trouble is, surprisingly enough, computers. During 1962-63 work was computerised, but at the crucial period of changeover to punched cards the staff was allowed to dwindle, just when it should have been augmented. The planning of the new system was inadequate and the division of responsibility not defined!'

to accept that the computer assisted their work or had advantages that outweighed the apparent disadvantages, the majority adapted themselves to the new methods of work.

Relationships with the three subsequent branches to be taken over were easier as many of the early difficulties had now been ironed out, but there were still some problems. A complaint at all branches was that people were not kept sufficiently in the picture by the computer centre. The staff of the fourth branch to be transferred complained as a group about the kind of information they received; 'we were given screeds and screeds of written notes on the new system but this meant nothing to us. What we really needed was to spend a period of time at a branch that had already been taken over and see for ourselves what was going to happen.' This branch also found it difficult to talk to computer centre staff. The accountant said 'they will keep dragging in technical jargon and it's just like gibberish to us.'

Branches also complained about the number of false alarms they were given by the computer centre concerning the date of transfer. They told the research team that they were given a date and then it was later cancelled and the change-over postponed. These delays were, of course, due to the serious technical problems that were being encountered at the computer centre but branch staff were not always aware of these. Had they been, their reactions might have been more sympathetic. One manager suggested that the computer staff deliberately kept quiet about their difficulties in order to reassure the branches that there would not be problems and that branch staff had nothing to fear. Therefore, they played down the troubles by trying to pretend they did not exist. The manager thought that this policy was mistaken and believed it would have been easier for the branches if the computer staff had treated them with complete frankness. The appointment of a computer centre manager in October 1963 somewhat eased this problem. He was at the Centre all day and this meant that a branch manager could contact him about any unexpected problem. The O. & M. Manager later told us that branch managers could always have contacted his own department with their queries but at the time they did not appear to appreciate this fact.

Relations with the computer manufacturers.

In common with many other firms installing their first computers at this time, a certain amount of conflict and disagreement developed between the computer personnel and the computer manufacturer.[6] This was due partly to the manufacturer's inability to meet promised delivery dates and partly to the succession of breakdowns affecting both the computer and ancillary equipment. Throughout the installation period progress was continually hampered by late deliveries of vital equipment. This was true not only of the computer itself, but also of the ancillary equipment both at the Centre and the branches and, in its turn, of the line-printer. The delivery date for the latter was continually postponed until it eventually arrived in October 1962.

These breakdowns not only irritated and inconvenienced the Centre staff by causing them to work very late, but also caused staff to worry in case senior management misinterpreted the situation and concluded they were not competent to run the computer. The programmers felt that, as a result of the continual breakdowns which the manufacturer's engineers failed to eliminate, they personally might be held responsible for a lack of progress which was to a large extent outside their control. This lack of progress did, in fact, cause a deterioration in relationships between the Centre and the branches. It seemed that, as might be expected, the bank was initially over-optimistic in its estimate of the time required to change over from manual data processing to E.D.P. Senior management then became surprised and worried when time schedules were not met. The computer staff also complained that it was difficult to make the manufacturer's resident engineers appreciate the need for absolute accuracy in banking, as this kind of work was entirely new to them.

A representative of the computer manufacturer later admitted that his firm had learnt a great deal from introducing this particular computer and that they would have to take more care in future with 'bank jobs'. It was not like installing a university computer where time was not important; the bank must meet time schedules and the manufacturers had not sufficiently realised this.

[6] We found this at our second case study firm and also at other firms visited during the course of the research.

As time went on machine faults were corrected and breakdowns became fewer but, as one of the programmers pointed out, in the early days of the installation the Centre staff were, in effect, doing research and development for the manufacturers of computer equipment. Manufacturers were able to modify and improve equipment on the basis of the bank's experience. Although this was undoubtedly of great benefit to the manufacturers it was both time wasting and frustrating for the computer centre staff. Despite improvements, as late as March 1963 the programming staff were not entirely happy with the technical reliability of the computer. One programmer suggested at that time that it was imperative to get the computer working really reliably as soon as possible.[7] The Centre was now reaching a stage when it could no longer afford to have breakdowns as there just was not the time available to put them right.

The Royal Exchange Bank, of course, suffered from its having entered the computer field at such an early stage. Commercial computers were comparatively new in 1961 and manufacturers did not, at that time, completely appreciate the requirements of business organisations. The bank's senior programmer maintained that had they been able to wait another 12 or 18 months before getting their computer they could have obtained a more reliable machine, and one that had greater adaptability. He felt that the Centre had learnt a great deal from the problems associated with making the first computer work, and with future computers they would, for example, be very careful to have an agreement with the manufacturer so that there would be some redress in case of continual breakdown. The opinion seemed to be generally held by the programmers that an inadequately tested machine had been supplied to the bank.

[7] Soon after the computer became technically reliable, it also became obsolete. Far better computers were now on the market.

CHAPTER 4

The Royal Exchange Bank—Work organisation and staff attitudes before the change

This chapter describes the situation in the branches before the computer changed the system of work. We will first consider the kinds of people the bank clerks were, their attitudes to their jobs and their ambitions for the future. We can then try to establish whether these personal characteristics and career objectives influenced their reactions to the 'idea' of office automation. If bank clerks are a special kind of group quite different from, for example, coal miners or research chemists, this means that management will need to adopt policies for change that do not conflict with the particular kinds of satisfaction they are looking for in their jobs.

The work of a branch

In order to understand what the bank clerks are referring to when they comment on their jobs it is necessary to know something of the routine work handled by a branch. Here is a brief description of this.

Routine branch accountancy can be divided into two principal functions—handling credits and handling debits. Credits are either cash and cheques passed into the bank over the counter by customers, or credit notes and cheques coming through the post from other branches or banks. Debits are cash handed across the counter in return for customers' cheques, or money paid out of customers' accounts to meet cheques drawn on the bank. Both credits and debits have to be 'cleared' at the branch; that is, the amounts must be listed and the cheques scrutinised to see that words and figures tally and that the customers' signatures are genuine.

Once a debit cheque is paid or a credit received these credits and debits pass through the 'waste'. This is a checking process to ensure that each credit item, except cash, has a corresponding debit item. For example, cheques paid in but drawn on other banks will be recorded as a credit to the receiving bank and a debit to the paying bank. (The total of cash paid in is balanced with the cashier's books after the counter has closed in the afternoon.)

These transactions are then ledger posted. This means that the amount paid out or in is listed on the customer's account record and a new balance worked out. Ledger posting is followed by statement posting; now the same credits, debits and balances are listed on customer's statements so that a customer can be given an up to date statement of his account whenever he requests this.

A final clerical operation is 'journals'. This is the gathering together of all totals produced by the different bank sections, including both machined totals and the totals of odd accounts which have been dealt with manually. These section totals produce a final total of all debits and credits handled by the bank that day. All totals are gathered together in a 'journals summary' which is then entered into the branch's general ledger.

In order to cope with these operations and with its customers, the bank employs several different categories of staff. At the top of the branch hierarchy are the manager and assistant manager who, among other things, look after customer relations and deal with requests for loans. Next comes the accountant who is responsible for the routine day to day supervision and operation of the office. Cashiers handle cash and cheques at the counter and are in a face to face relationship with the customers. Both women and men are employed on this job, although the employment of women is a recent development. In the office there are ledger controllers—also known as control clerks— who are responsible for the payment of cheques and for establishing that these cheques have been correctly completed by the customer; this tends to be a man's job although some women are employed on it. Girl machinists post the ledgers and statements and look after the waste; a journals clerk balances the end of day totals, and there are a number of general duty clerks, 'spare men' and juniors who perform miscellaneous jobs and take over when a member of staff is absent.[1]

[1] This is a very brief and over simplified description. It must be emphasised that most branches deal with a much broader range of functions than those noted here and employ a number of specialist staff, such as security clerks.

The five branches and their managers

Royal Exchange planned, in the first instance, to process the work of five of its central branches on the computer. In order of proposed date of transfer these were King George's Place, St. John's Avenue, Paul Street, Ringer Square and North Office. All were situated in the business area of the city and all were within a mile of the bank's new computer centre.

King George's Place was a medium sized branch with a staff of 23 and was physically closest to the computer centre. Like the other four branches, its work was predominantly commercial and it handled the accounts of some very large local firms. The research team interviewed staff there in August 1960, over twelve months before the computer took over branch work. The difficulties of this branch have already been referred to in the previous chapter and at the time of the pre-change interviews we found that both management and clerks had very little precise information on what was going to happen—although this lack of knowledge was common to everyone else in the bank. The manager, like the managers of the other branches scheduled for transfer, was unenthusiastic about automated banking. He did not wish to be bothered with the upheaval arising out of the introduction of a new system. The clerks had a small amount of general information about the computer but had few facts and were confused about what the new machine would do and how they and the branch would be affected. Computer centre staff were already working in the branch, personalising customers' cheques, but they had little information themselves on how the new system would work in practice. We were to find that this branch's unfortunate early experiences of office automation had a considerable influence on the reactions of the clerks in the other branches, and led to a great deal of alarm and anxiety.

The second branch, St. John's Avenue, was very much smaller than King George's Place. Branch business was handled by a staff of 15, including the manager and assistant manager. Interviewing was carried out here in December 1961 and we found that this manager, like the previous one, was unhappy about the computer and appeared to have little precise information on the programme of events. He and his staff had known unofficially for some time about the change over to E.D.P., but had only been informed officially in the preceding January, shortly after cheque personalisation had commenced in the branch. They still had no date for the final transfer of work.

The third branch, Paul Street, was larger with a staff of 49. A considerable proportion of its work was managing the accounts of one major local commercial firm. The research team went to Paul Street in January 1962, when the problems at King George's Place were at their height. They found that this manager, although worried about the future and his position, was prepared to accept the idea of office automation, believing that it had to come some time. His attitude was that 'the change has got to be made but it must be made as smoothly as possible'. His staff, however, were becoming increasingly apprehensive; 'the girls keep asking me what is going to happen' said the manager, 'but I can't tell them as I don't know.'

Ringer Square had a staff of 23 and was rather different from the other branches in that its business was less commercial. It handled the affairs of many old established families who, even though they had moved away from the district, still preferred to be looked after by this particular branch. Interviewing took place here in March 1962. Both management and staff appeared less apprehensive than those in other branches, partly because their time for transfer to the computer was a long way ahead, partly because this branch was the furthest distance from the computer centre and some distance from the other branches, and so was isolated from the flow of rumour.

The fifth and final branch, North Office, with a staff of 109, was very much larger than the others—in fact it was the bank's biggest office. The last of the pre-change interviewing was completed here in April 1962. The North Office manager, like the previous one, was not as worried about automation as the earlier managers and, as it turned out, he had no cause to be, for a late change of plan resulted in only a very small section of the work of his branch being transferred to the computer.

At this early stage of the research, we were left with a number of impressions which were later checked against our interview data. It seemed that in all the branches there was a lack of specific information about the computer and particularly about the phasing and timing of events—and that managers were as much in the dark here as their staff. This lack of information appeared to have a number of causes. It was partly a result of the physical separation of branches from each other and from the computer centre, which impeded the accurate transmission of information, partly due to an inadequate formal communication system and insufficient consultation with branch managers, and partly a consequence of the experimental nature of the whole computer operation which made exact planning impossible.

Royal Exchange had made some attempts to keep its staff informed of computer developments but these did not prove very successful. Early in 1960 there had been a staff circular to branches and the bank Staff Association was also informed at this time.[2] The possibility of a computer had been mentioned every year at the staff dinner and there had been a number of lectures by the manager of the Organisation and Methods Department including one given at the request of the Staff Association. Articles were published in journals likely to be read by bank staff and there were notices in the local and national press. Many of these articles explained in considerable detail the plan for the computer so that, for those who read them, very full information was made available. In fact, of course, it is very difficult to convey information of this kind, and the analysis of interview data showed us that few of the clerks had more than a general idea of what the computer was to be used for, although they all knew that such a machine was coming.

During this pre-change period it seemed that informal communication was far more effective than formal. The difficulties encountered by the first branch to go over to the new system were soon common knowledge and aroused a great deal of alarm in the branches next in line for transfer. Stories circulated about machine breakdown, continual late working and staff going off with nervous breakdowns. The managers became worried about the effect of the computer on their own positions and on customer relations. They were disturbed at the idea of detailed account information being removed from the branch to the computer centre. With the traditional system, the manager had an up-to-date record of every transaction a customer made and could, for example, spot immediately if a customer was incurring a large number of hire purchase debts—a factor which might affect his

[2] The National Union of Bank Employees was not recognised by this bank.

willingness to grant an overdraft. With the new computer system the manager would receive a daily note of a customer's balance, but he would no longer have detailed information immediately to hand. Again, although customers could obtain statements whenever they wished provided notice was given to the computer centre, a customer could no longer walk into a branch and demand a statement then and there.

This brief outline of managerial reactions to the idea of a computer illustrates the anxiety most people feel when presented with an entirely new situation. The Royal Exchange managers had banking at their fingertips and were confident of their own knowledge and ability in the traditional banking situation. A radical change in this was seen as presenting them with all kinds of problems and uncertainties. They appreciated that the new system's speed and efficiency would benefit the bank as a whole, but at this early stage of its adoption they were not convinced that it would have any advantages for them personally. Any bank installing a computer today would not, of course, have this long waiting period between the transfer of work of individual branches. But, given the same absence of specific information as we found in Royal Exchange, there is no reason to suppose its managers would be any less anxious.

The bank clerks[3]

In each branch we interviewed all those clerks whose work was likely to be altered by the computer. This gave us a total number of 101, 36 men and 65 women. Although at this early stage we could not be completely certain which jobs would be affected, it seemed probable that clerks dealing with routine accountancy functions would feel the impact of the computer most of all. These were the control clerks, clerks engaged on general duties and the ledger posting machinists. We interviewed all clerks occupying these jobs, plus any person who thought his work would be changed by the new system and a few clerks who had been made responsible for its introduction. Cashiers, typists and clerks on specialist work, for example, securities, were thought unlikely to be affected and they were not interviewed. Because clerks on

[3] Unless otherwise stated all statistical correlations will be significant at least at the 5% level.

routine duties are usually at an early stage of their careers, our criteria for selection provided us with a young interviewee group. 83% were 25 years or less, with the girls rather younger than the men.

SOCIAL CHARACTERISTICS

Bank clerks are always regarded as very much a middle-class group and in fact this proved to be the case with our sample. We found all but 5% of the male bank clerks came from middle-class homes and had fathers who followed some kind of white-collar occupations. They had been to grammar, independent or public schools and passed their G.C.E. 'O' levels, although not necessarily at a high level or in many subjects. Nearly all had joined the bank immediately on leaving school and, once there, they continued with their studies, working in the evening for the Certificates of the Institute of Bankers, a necessary preliminary to promotion. Most of them expected to stay with the one bank for the whole of their working lives, an interesting exception to the modern practice of spiralling from firm to firm in order to progress up the management ladder. We asked why they had decided to become bank clerks and found that a third had been positively interested in the work, a third had been influenced in their choice of job by a relative or friend, and a sixth had been attracted by what they believed to be favourable conditions of work. The remainder gave negative reasons such as a lack of training for anything else. It was unusual for any of these clerks to imply that banking had been a 'second best' choice of job.

The women bank clerks, as a group, were rather different from the men. 21% came from working-class homes, less than a half had attended a grammar school and only a half had G.C.E. at 'O' level. In contrast to the male clerks, many had taken a period of further full-time education after leaving school, usually in shorthand and typing or in commercial subjects. This further education was related to the demands the bank made of women entrants. The tradition was to take grammar school girls and it seemed that secondary modern school girls were normally only accepted if they had a special qualification, such as proficiency in shorthand and typing. This was accepted as a substitute for the G.C.E. of the grammar school girl. We found that this shorthand typing requirement was a source of frustration to many of the girls.

Having acquired the necessary skill they were never in a position to use it, for a small number of girls could cope with all the typing that branch accountancy required. Most women joined the bank on leaving school and the majority remained for several years, although few stayed long after marriage. There were not many career women in our five branches and none had a position higher than supervisor of female staff.

We found that although nearly a quarter were of working-class origin, the girls appeared to think that in joining the bank they had also joined the middle class. When asked to describe their own social class position only 4% replied 'working class'.

ATTITUDES TO THE OLD SYSTEM OF WORK

In the following analysis of clerks' attitudes, the four branches which went onto the computer will sometimes need to be distinguished from the fifth, North Office, where only a small part of the work was transferred. This will facilitate the comparison of pre- and post-change reactions to office automation. So that the reader may be clear which branches we are discussing, the four automated branches will be called the 'E.D.P. branches'.

In all five branches there was a clear-cut division of work between the sexes, with the men generally working on the higher status and more responsible jobs and the women on tasks that were simpler or involved the use of accountancy machines. The table below shows the jobs of the clerks we interviewed and the numbers in our sample employed on each job.

TABLE 1 *Jobs of clerks who were interviewed*

	MEN	WOMEN
Control clerks	12	3
General clerks and spare men	9	3
Juniors	5	—
Miscellaneous jobs	3	7
Machinists	—	45
Supervisors	5	5
Staff preparing for the new computer system.	2	2
TOTAL	36	65

TABLE 2 *Attitudes to work*

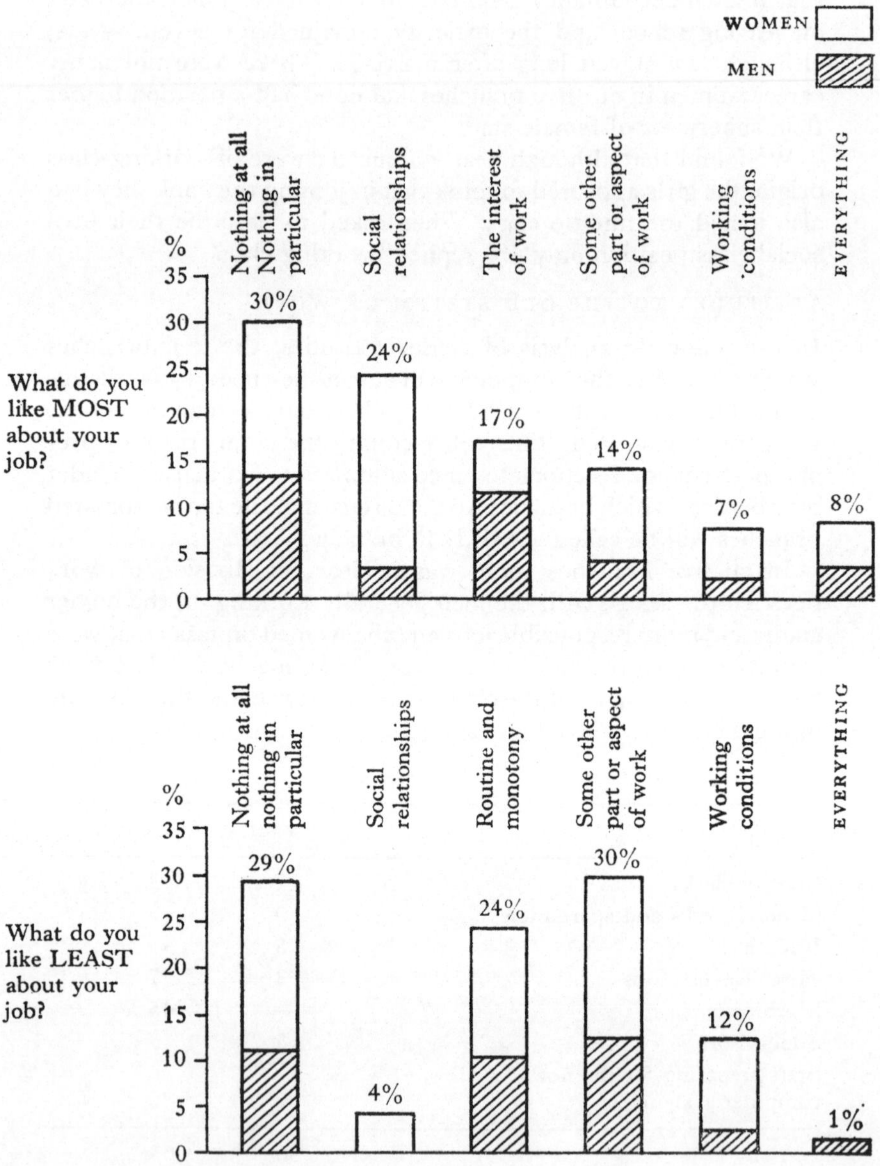

We asked the clerks what they liked most and least about their jobs, hoping to achieve two things by these questions. First, to find out what bank clerks considered to be the attractive and unattractive features of their work. Second, to obtain comparative information which would enable us to check whether they gave the same replies to these questions when we interviewed them a second time after the computer had taken over.

There was little unanimity in the clerks' replies and it seemed that there was no one feature of bank work that appealed equally to everyone. Indeed, the answers showed that many clerks had a rather negative approach to their jobs—over a quarter were unable to think of anything they especially liked or disliked. Only 17% referred to the 'interest' of their work, a response we noted carefully as work interest was something the computer might very well affect.

Men and women tended to have similar opinions although men were more inclined to speak approvingly of the interest of their work and women of pleasant social relationships. Three-quarters of the women giving this last reply worked in North Office and it seemed that there were factors operating in this branch that aided social interaction—at least amongst women. Two rather obvious ones were the large number of women employed there—we interviewed 32—and the fact that most of the tasks involved an association with another person.

The clerks' answers to these two questions suggested that North Office men were less contented than men in the E.D.P. branches. Nearly a half were unable to suggest anything they liked about their work. Again there appeared to be reasons for this negative attitude. The men were less interested than the women in social relationships as an end in themselves, and the size of the branch was disadvantageous from their point of view as it led to narrow job specialization and small routine jobs. Also the research team were told by some of the young men at North Office that clerks who were not progressing as well as they should were sent to this branch so that senior management could keep an eye on them.

If we consider the answers of the clerks in terms of the jobs they were doing, one other noteworthy point emerges. All the women saying that routine and monotony were the aspects of work they liked least, were machinists. Their job was both demanding and dull for they had to concentrate if they were to avoid mistakes, they could not talk as they worked, and their task was a simple listing of data from cheques and vouchers. We wondered if the work of these girls would become even more routine and dull once the computer took over.

JOB SATISFACTION

The answers to the questions discussed above suggested that some groups were more satisfied with their work than others and it seemed important to try to identify these groups more accurately. After the change we would want to know whether job satisfaction increased or decreased and whether any shift was general or applied only to certain people. However, we realised that assessing job satisfaction is not an easy thing to do. People's attitudes to their jobs are rarely simple and straightforward. They vary from one moment in time to another and are affected both by incidents at work and by factors in the individual's social or home environment. Commonly used measures of job satisfaction are labour turnover and absenteeism—indicating a desire to withdraw from an uncongenial situation—and the prevalence of complaints, grievances and strikes. Many of these measures are crude, inaccurate and often meaningless and would be inappropriate for our purposes as well as unobtainable. Absenteeism and labour turnover amongst young women often have nothing at all to do with job satisfaction, whilst once men enter a bank they are largely committed for life and to leave would mean changing their careers. The bank as yet has no formal machinery for the handling of day-to-day grievances and so far there has not been a strike in the history of the clearing banks.

Despite these difficulties, our pre-change questionnaire included a number of questions which seemed to provide indications of degrees of job satisfaction. We asked the clerks, if they were starting their careers again would they go into a bank? Would they like their sons to become bank clerks? What were their promotion prospects, and the question already analysed in Table 2, 'what do you like most about your job?' It seemed likely that clerks who gave affirmative answers to the first two questions, who believed their career prospects to be good and who said that what they liked most about their jobs was the interest of the work, would feel high job satisfaction.

Although we appreciate that this is a rough measure of job

TABLE 3 *Indices of job satisfaction*

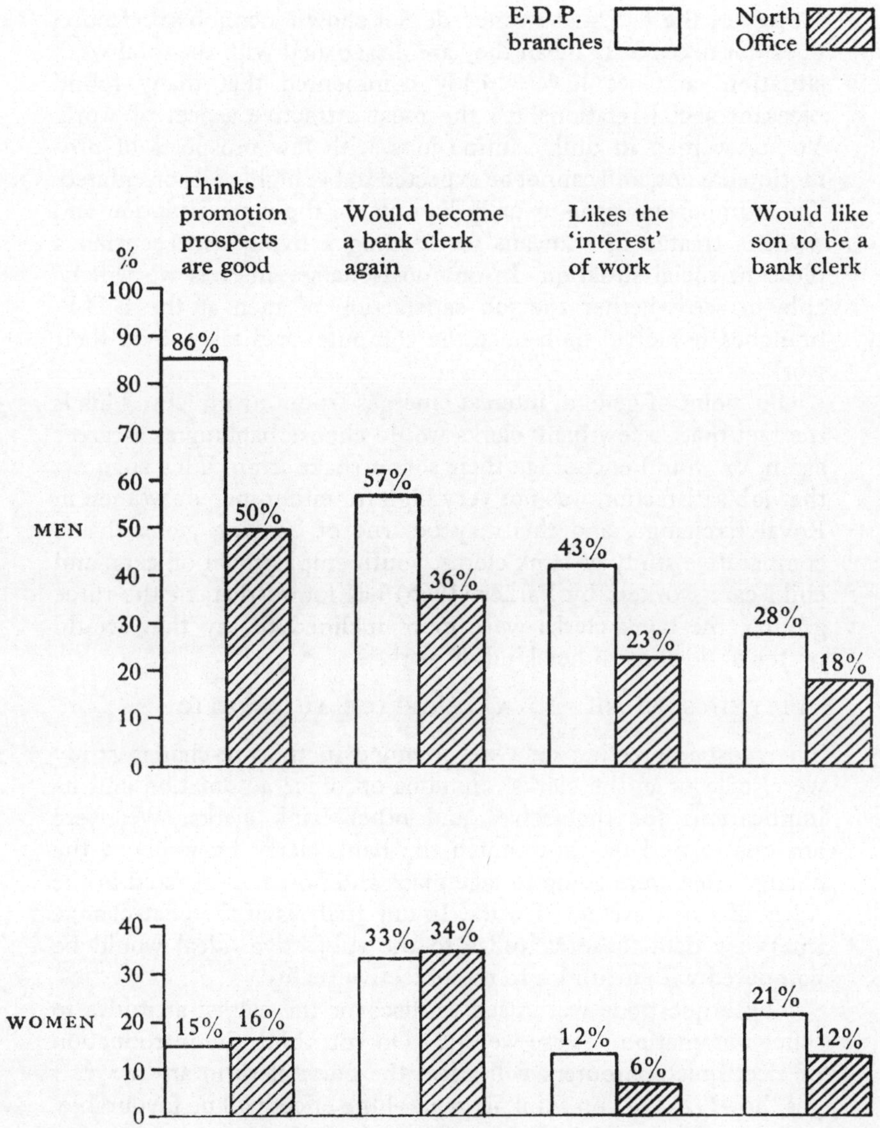

satisfaction, our indices again suggested that men at North Office were less satisfied with their jobs than men at the E.D.P. branches, while women showed the lowest degree of job satisfaction. However, the fact that women do not show much job satisfaction does not necessarily mean they are dissatisfied with the total work situation, and we have already commented that many found pleasant social relationships the most attractive aspect of work. Young women in dull, routine jobs with few prospects of promotion are not, and cannot be expected to be, highly job-orientated. Their important goals usually lie outside the work situation and work is treated as a means for earning a living and securing a pleasant social situation. In our post-change analysis we will be able to see whether the job satisfaction of men at the E.D.P. branches remained high once the computer had taken over their work.

One point of general interest emerges from our indices. This is the fact that so few bank clerks would choose banking as a career again, or would encourage their sons to take it up. This suggests that job satisfaction was not very high for either men or women in Royal Exchange, and this may be true of banking generally. A comparative study of bank clerks, youth employment officers, and child care workers by Parker (1965) has found that, of the three groups, the bank clerks were more inclined to say they would prefer to do some other kind of work.

ATTITUDES TO THE IDEA OF OFFICE AUTOMATION

The most interesting data we obtained in the pre-change study were, of course, the clerks' opinions on office automation and its implications for themselves and other bank clerks. We were anxious to find out how much the bank clerks knew about the changes that were going to take place and how they reacted to the 'idea' of a new system of work. In our analysis of the post-change interview data these attitudes to the subjective 'idea' would be compared with attitudes to the objective reality.

Three questions were used to discover the clerks' attitudes to office automation. These were (1) Do you think the introduction of electronic computers will affect the bank clerk in any way? If yes, how? (2) Do you think in general the effect will be favourable, unfavourable or neither one nor the other? (3) Do you think that you personally will be affected? If yes, how?

TABLE 4 *Do you think the introduction of computers will affect the bank clerk in any way? If yes, how?*

	YES	YES My work will become					YES	NO Don't know	TOTAL
	Staff numbers will be affected	More interesting	Less interesting	Easier	Harder	Different in other ways	Customers will be affected		
MEN	6	8	2	3	—	10	3	4	36
WOMEN	17	1	12	5	5	12	4	9	65
	23	9	14	8	5	22	7	13	101

Nearly all the clerks believed the bank clerk would be affected in some way by the introduction of computers, although they were not clear what form these effects would take. The majority spoke of changes in work—either work organisation or work content, a quarter believed computers would alter the number and kind of clerks required by banks, while a few clerks suggested that in the future the bank would have a different kind of relationship with its customers.

Although clerks were not asked in this first question to evaluate the changes they foresaw, nearly a half spontaneously did so, showing that there was a clear cut division of opinion between the sexes on whether these changes would have favourable or unfavourable consequences.

TABLE 5 *Opinion on changes shown in Table* 4

	FAVOURABLE	UNFAVOURABLE	TOTAL
MEN	13	4	17
WOMEN	7	23	30
	20	27	47

Women, on the whole, believed that work would become less interesting and their comments were both critical and derogatory.

> 'Yes work is going to be very monotonous for everyone.'
> 'We will all become like factory workers—watching machines and feeding them.'

Men, on the other hand, saw computers as making work more interesting.

> 'The computer will take a lot of monotonous work away, but more from machinists than from ourselves.'
> 'Routine work will be eliminated. There will be more specialised work for men.'

They also saw fundamental changes taking place in branch organisation. One clerk told us:

> 'In the long term the computer will change banking completely. The average branch will be a manager, a clerk and a machinist.'

The popular idea of a computer's consequences is lines of clerks queuing up outside employment exchanges. We were anxious to know whether those bank clerks who spoke of staffing changes would see this picture. In actuality, they did not. Most women were certain that girls would be displaced by the computer, but they saw this displacement leading more to the transfer of staff to other branches, than to redundancy. The men too spoke of displacement rather than redundancy, although they were less certain who would be affected. Some suggested that there would be fewer men and more girls, others that there would be fewer staff altogether.

This lack of anxiety about redundancy was almost certainly a product of the clerks' banking environment. Bank clerks have always known security and the majority supposed that it would continue.

We mentioned earlier that a few clerks suggested that the bank's relations with its customers would change, and the opinion was almost unanimous that such a change would be for the worse. They said: 'The old friendliness will go; the customers won't like it.' This was the first indication of an alarm about one consequence of the new system which was to grow and spread as the change-over progressed.

It can be seen that the clerks' impressions of how the computer would affect their jobs were rather vague. Women saw a blacker picture than men—a factory type environment full of machines, a reduced need for women staff, and an increase in monotonous work. Men predicted general organisational changes—although they were uncertain about their nature—but they also saw the computer as eliminating routine and monotony.

The follow-up question, asking the clerks whether, in general, they thought the effect of the computer would be favourable or otherwise, produced the following results.

TABLE 6 *Do you think, in general, the effect (of the computer) will be favourable, unfavourable, or neither one nor the other?*

	FAVOURABLE	UNFAVOURABLE	NEITHER	BOTH	Don't know	TOTAL
MEN	21	3	2	8	2	36
WOMEN	31	10	4	8	12	65
	52	13	6	16	14	101

There was no significant difference between the answers of men and women, whereas spontaneous comment to the previous question had shown a clear cut division of opinion on whether the computer would have favourable or unfavourable consequences for bank clerks. This could have been because the second question contained the phrase 'in general' suggesting to the clerks that they should answer in terms of bank objectives rather than their own immediate interests. There was also some indication that clerks were now taking a longer term view.

> 'In the long term it will make the place more efficient'.
> 'Favourable in the future, although before its introduction there will be a lot of work for some people.'

Unfavourable responses came largely from the Paul Street branch, where over a third of the staff were antagonistic to the idea of a computer. The reason for this is almost certainly to be found in the timing of the interview programme. We interviewed at Paul Street when King George's Place, the first branch to have its accounts transferred to the computer, was still experiencing very considerable difficulties of adjustment. In particular, breakdowns at the computer centre were causing hold-ups at the branch and involving the staff in a great deal of late work. There were also problems connected with the branches' new mechanical equipment. Rumours of these difficulties had reached Paul Street branch and not unnaturally were causing anxiety, especially as they were expecting their own branch to be involved very shortly.

The last of the three questions on change was intended to find out how the clerks thought the arrival of a computer would affect them personally. In reply, 54 said they would be affected, in contrast to the 88 saying 'yes' to the more general question on whether automation would affect bank clerks. As our group of clerks was chosen because these were the individuals most likely to have their jobs changed by the new system, it was surprising that 47 were not convinced that this would happen. There was no significant difference between the sexes in the frequency with which the clerks said 'yes' they personally would be affected, another surprising result since it is the job of the female clerk which is, in fact, most likely to be altered by office automation. However, we did take some pains to single out for interviewing

TABLE 7 *Do you think that you personally will be affected? If yes, how?*

	YES	YES My work will become:					NO Don't know	TOTAL
	I shall be redundant or transferred to another branch	More interesting	Less interesting	Easier	Harder	Different in other ways		
MEN	3	3	1	—	2	9	18	36
WOMEN	11	—	1	3	2	19	29	65
	14	3	2	3	4	28	47	101

those male clerks most likely to feel the impact of the computer, and this no doubt influenced our results.

Table 7 shows that when clerks were asked the more personal question, those that answered at all tended to give specific answers usually related to changes in the organisation or content of work, such as 'we won't be on the present machines', and there was a decline in the numbers of these answering in general but more qualitative terms (e.g. 'more interesting work', 'easier work'). This suggests that although the clerks knew that their work would be organised differently after the computer, they did not know what the change would imply in qualitative terms for them as individuals.

FACTORS INFLUENCING THE CLERKS' ATTITUDES

The next step is to try and identify some of the reasons for the different replies from the clerks. Why did they not all view the consequences of office automation in the same way? Why did some believe a computer would lead to advantageous changes while others saw only problems? Could there be factors in the clerks' backgrounds, in their aspirations or in their work situations that influenced their attitudes to change? An American study of office automation carried out by a research team from Michigan State University has identified a number of characteristics which appear to influence receptivity to change (Faunce 1960). These are class background, education and job status. Let us consider these first of all and then see if any other variables appear to have influenced the attitudes of our clerks. The clerks' spontaneous evaluation of the favourable or unfavourable consequences of office automation when they answered the question 'Do you think the introduction of electronic computers will affect the bank clerk in any way, if "yes", how?' will be used as one index of attitudes to change. Answers to the question 'Do you think in general the effect will be favourable, unfavourable, or neither one nor the other?' will be used as another index. We will see whether, for example, clerks from working class backgrounds tend to answer in a similar way while clerks from middle-class homes have quite different opinions.

The American team did find a relationship between class background and a willingness to accept change, with the higher class status groups more pro-change than the lower. They also found that higher educated groups were more in favour of change than lower ones. Their conclusion was that a higher social class position and a higher education go with a greater willingness to accept change. In our analysis we were unable to find any relationship between either class background or the kind of school attended and attitudes to the idea of office automation. We did, however, find a relationship in the opposite direction, for girls with G.C.E. were more likely to see the change as unfavourable than those with no academic qualifications. Although this contradicted the American findings it seemed a rational reaction. These girls had entered the bank with good scholastic qualifications and probably higher aspirations than the unqualified or lesser qualified girls, and they therefore had more to lose from a system which increased the mechanical and routine nature of work.

TABLE 8 *Scholastic qualifications of women clerks by their opinions of office automation*

	FAVOURABLE	NEUTRAL	UNFAVOURABLE	Don't know	TOTAL
G.C.E.	4	12	16	—	32
Commercial qualifications	—	5	6	5	16
No qualifications	3	9	1	4	17
	7	26	23	9	65

The American team found too that clerks in higher status jobs were more pro-change than clerks in lower status jobs, but here again our analysis did not provide any convincing supportive evidence. Certainly the male bank clerks were more favourably disposed towards change than the women and they tended to be in the higher status jobs. However, it seemed likely that their favourable attitudes were due more to the fact that they believed themselves to be less threatened by the change than to their position in the status hierarchy.

Although the social class, education and job status of the bank clerks could not be related to their attitudes to the proposed change, there were a number of other factors which did appear to be related. First, girls who had always worked in the same branch of the bank were far more pro-change than those who had moved around from branch to branch. These girls were nearly all located in North Office and their attitude suggested a reaction against boredom; change being seen as a means for relieving a monotonous work routine.[4]

Second, men who believed their promotion chances to be good were very likely to see the general effects of automation as favourable. These men believed that their future was secure. They had reasonably high job satisfaction, they were identified with the bank's interests, and had confidence in bank policy, and they did not feel threatened by the computer. As one confidently said 'it must be the right thing to do. I have a high opinion of bank planning. They wouldn't overthrow the present system without a better one'.

It seemed that the most significant factor conditioning the bank clerks' attitudes to the idea of a computer was how they saw the change as affecting themselves. Girls, particularly those with scholastic qualifications, saw it as damaging to the interest of work, and only those who were bored enough to welcome any change thought it a good idea. Men on the other hand believed they would not be affected very much by the change and, providing their career prospects were good, they were prepared to welcome it as a means for improving the efficiency of the bank.

Only one variable was common to both men and women and this was related not to the perceived effects of the change, but to the way information about the change was obtained. There was a strong tendency for clerks who said they had heard about the computer from unofficial sources, such as friends or rumour, to be less favourable in their attitudes than those who heard officially from management.

TABLE 9 *Attitudes to automation by source of information*[5]

	FAVOURABLE	Neutral, ambivalent, don't know	UNFAVOURABLE	TOTAL
Heard officially	23	10	2	35
Heard unofficially	29	20	11	60
	52	30	13	95

[4] Girls on routine jobs seem to try and introduce variety in different ways. Young women factory workers in some areas are known to move around from firm to firm, staying only a short time in each.

[5] This table excludes six clerks who said automation would have no effect.

We were surprised to find that so few clerks remembered hearing about the computer from official sources, despite the early attempts the bank had made to inform staff. Less than a third told us they had heard about the computer in 1960, when the official staff circular went out. The majority claimed they had not heard anything until 1961, when the arrival of the computer apparently acted as a stimulus to the spread of information through the bank grapevine. The replies of the clerks tell us, of course, only when they believed they first heard about the computer, not when they did, in fact, first hear. Clerks may have been told officially and forgotten this or the information may not have registered in the first instance.

The following table shows that the effect of official communication diminished as the change progressed, and rumour gradually took its place.

TABLE 10 *Percentage of clerks in each branch hearing about the computer from official or unofficial sources*

Branch in order of proposed transfer to the computer.	Official sources %	Unofficial sources %
King George's Place	78	22
St. John's Avenue	75	25
Paul Street	36	64
Ringer Square	42	58
North Office	28	72
All clerks	37	63

There were two apparent reasons for informal communication replacing formal as the principal source of information. First, the restriction of official communication to the early circular meant there was no mechanism for providing the clerks with continuous information; they had to rely on the grapevine. Second, although clerks in the later branches to be transferred did receive official information shortly before the change-over took place, obviously they had plenty of time to hear about the new system unofficially from friends in branches already transferred.

The question of communication and the introduction of

technical change will be discussed in some detail in Chapter 10. Our pre-change analysis does, however, bring out two important points. These are the ineffectiveness of a small amount of official communication put out in advance of an innovation, and the greater potency of unofficial media such as talk and gossip as a means for spreading information.

CONCLUSIONS

This analysis of pre-change opinion in Royal Exchange is useful for a number of reasons. It provides us with information on the clerks' attitudes to their jobs and thus allows us to see if these attitudes change once the computer has been introduced. Although it will not be possible to say with absolute confidence that any change in attitude is due to the computer, if a similar change does not occur in North Office, where the computer's impact was minimal, this will suggest that there is a relationship. Pre-change opinion also provides us with the clerks' reactions to the idea of office automation and enables us to see if they still hold the same views when faced with the reality. Lastly, our attempt to identify variables influencing attitudes to change has been productive to some extent. In the post-change analysis we will see if the same variables still influence attitudes or if there are now different or additional variables.

We can now make one of the points that we shall be stressing throughout the book. This is that if different groups or individuals have varying perceptions of how a particular technical change is going to affect them, and react favourably or unfavourably because of these, then management needs to know what these perceptions and reactions are so that it can plan its change policies to fit the human relations needs of the situation.

CHAPTER 5

The Royal Exchange Bank—The new work system

In this chapter we look at the branches after the change-over to electronic data processing, seeing what the bank clerks thought about the new system of work and how they believed it had affected both themselves as individuals and the bank as a whole. Wherever possible the opinions of clerks in the E.D.P. branches will be compared with those of the clerks in North Office. This will provide us with a rough control group.

It will be recalled that although North Office had been scheduled for transfer to the computer, and we had completed our pre-change interviews there in the expectation that this would happen; in practice only a small fraction of the work was ever automated. The bank clerks' personal accounts were processed electronically, but customer accounts were still handled in the traditional way. The impact of the computer was, therefore, very slight and North Office can be regarded as a virtually unaffected branch. Despite this, the attitudes of staff there can provide us with some interesting additional information. In the first place we can establish the post-change attitudes to automation of staff in a non-affected branch and see if these are the same as the attitudes of staff in affected branches. If not, to what extent and in what areas do they differ? Secondly, we can establish the spread of communication in the bank and see how much North Office staff knew about the changes that had taken place in the branches, and how accurate a perception they had of the nature of these changes.

Once a branch was transferred to the computer, operations such as ledger and statement posting and the listing of daily transactions were completed automatically. This removed from branch machinists the work they had been accustomed to perform and also took away some of the control clerk's operations. Machinists were still required with the automated system but their job now was to punch financial data taken from cheques and credit notes onto tape for the computer.

We found that total branch numbers changed little after the computer took over, although the numbers of both machinists and control clerks were reduced. However, the bank believed it had achieved some saving through the new system of work for branch business had increased yet no new staff had needed to be engaged.

Managers as well as their staff were affected by the computer, for the old branch autonomy had now gone. Detailed records of customer accounts were no longer compiled by the branch and managers had to rely on the computer centre to provide them with this information. Let us start by making a short survey of managerial and supervisory opinion in each branch in turn, noting what were seen as the branch problems during and after the change.

Post-change problems in the E.D.P. branches

KING GEORGE'S PLACE

King George's Place branch encountered a number of problems in changing over. Increased pressure of work and difficulties arising out of machine breakdown have already been referred to in the last chapter, but the branch also met with resistance from customers, some of whom disliked the non-narrative statements which the bank had introduced simultaneously with the new computer system. When these customers were particularly valued and could not be persuaded to accept the new format, management gave in and produced narrative statements especially for them. This was possible as the branch still retained two of the old ledger and statement posting machines, but naturally it involved branch staff in a great deal of additional work, and it rather negated the advantage of having statements produced centrally at the computer centre. However, from a public relations angle this acquiescence with customer wishes was thought to be worth while and, in fact, the branch did not lose a single customer during the change-over. On the whole, senior management and office supervisors did not believe the new system would save them much work in the future. They found the preparation of data for the computer onerous and the branch accountant told us that automated methods would not prove really satisfactory until data could be read into the computer direct from the cheque. He thought the bank might have been well advised to postpone its automation plans until automatic cheque reading was established.

In this branch there was no saving in staff; while some jobs had been eliminated—for example the accounting process now had only to be completed once, instead of once for ledgers and once for statements—others had been created. A source of extra work arose out of the need to personalise and number all cheques and to number all documents. This had to be done in the branch and checked most carefully. If a document was not numbered the branch punch machines would not accept the entry. Dividends were a particular problem here. The bank had approached leading registrars in an attempt to persuade them to put customers' account numbers on dividends, but it was realised that this numbering at source would take a long time to organise.

Another new and time-consuming task was the need to trace errors that had caused the computer to reject data. If a tape was faulty the computer centre now returned it to the branch and it was the branch staff's responsibility to identify the mistake. Sometimes these errors were caused through mispunching by machinists, sometimes through faulty reproduction by the punch machines or misreading by the computer.

The supervisor of girls told us that she found supervision more difficult with the new system. With the pre-change procedure each girl had looked after the accounts of customers whose names came in a particular section of the alphabet—for example E.–G.—and she alone was responsible for all ledger posting and statements in this section. Now girls dealt with work as it was handed to them and it was virtually impossible for the supervisor to spot a girl who was slacking. Whereas the old system with its clearly defined allocation of work acted as a natural control mechanism, the new system required a closer supervision of job performance.

ST. JOHN'S AVENUE

St. John's Avenue branch was first visited after the change when staff were working on the half-yearly balance. It was the first time they had attempted this under the new system and branch management commented on how much this procedure was assisted by the computer. The accountant said that the computer had generally 'taken a lot of grind out of the day's work'. It was now possible to prepare in advance instructions for such things as standing orders as it was only necessary to punch a tape, and this helped to even out the work. He felt that branch staff must be watched, however, as there was a tendency to forget the new routines and operate on the basis of the old. One disadvantage was that branch work now seemed to be controlled by the computer centre, for deadlines imposed by the computer had to be met.[1] Another was the reduction in customer service. The accountant thought that customers had been 'put out' by the new system and he believed they disliked the fact that they could no longer walk into the branch and demand a statement. However, St. John's Avenue was more fortunate than King George's Place in that most of its customers were commercial firms and in consequence there were fewer problems with individual clients.

Staff at St. John's Avenue had been frightened of the new system at first, particularly when they heard of the problems at King George's Place—and management suggested that they should have been kept more in the picture by the computer centre. A supervisor commented:

> 'We were twelve months on the change. This could have been reduced if the computer centre had kept us informed. Certainly they had problems, but they did not seem to appreciate that these affected us. For a long time they kept saying "we're going to start the change-over next Monday", then they would ring us up and say "it's all off". Finally when the half yearly balance was due they said "we're coming now" and took the attitude "come on, what are you delaying for?"

Like King George's Place this branch had not been able to reduce its staff numbers, although a senior man had been replaced by a junior girl.

PAUL STREET

Paul Street also had an easier transition than King George's Place. The manager of this branch believed he had learnt a great deal

[1] This is an interesting consequence of introducing a computer. It brings the 'deadline,' which the shop floor worker has always known, to the office. However, a bank, which has always had to complete one day's work before commencing the next, would feel the impact of this time constraint less than the more laissez-faire commercial office.

from the unhappy experiences of the first branch to be transferred and he was determined to have everything highly organised before any of his work went on to the computer. Paul Street was fortunate in being able to move onto the computer in two stages. At first only ledgers were transferred while batch accounts were still completed under the old system.[2] A further factor slowing up the move to a complete E.D.P. system was the request from one of the branch's largest commercial customers that its accounts should still be handled in the traditional way. This firm was anxious to retain the facility of a daily statement itemising the previous day's transactions. The computer could not yet produce such a statement until noon the following day, while the customer required it at 9 a.m. This phasing of the change process meant that twelve months after the computer first impinged on branch accountancy, a very considerable part of the work was still processed under the old system.

Paul Street had 25 girls before the computer took over; afterwards these were reduced to 21. Business was continually increasing, however, so that more work was being done with less staff.

RINGER SQUARE

Work was not transferred at Ringer Square until April 1964, over a year later than the expected date. A new approach was adopted with this branch and the period of parallel running was abandoned. One day staff were operating the old system, the next day the new. This meant that a great deal of careful preparatory work had to be done and, in fact, preparations for the change-over were commenced 2½ years earlier. Unfortunately for the branch, continual delays meant that the re-organisation of procedure at the computer centre took place before the branch went on the computer. This led to many of the branch preliminaries having to be done twice over and made for a great deal of additional work and annoyance. Staff maintained that despite their early start at preparation, everything had to be re-checked in a great hurry immediately before the computer took over. In banking, this last minute rush when transferring information to a computer is inevitable.

[2] Batches or supplementary lists are used with accounts which contain a large number of different items. These items are listed separately, only the total figure appearing on the statement.

Customer account data is constantly changing and must be completely up to date when the computer goes into operation.

Managerial staff in this branch welcomed the computer even less than the management of the two previous branches for they were particularly concerned about the possible threat to good customer relations. Ringer Square was a very old established branch and still retained many customers whose families had done business with the branch for generations. These customers were elderly and had always received full narrative statements. The branch manager believed that this type of customer would not understand the new type of statement and might object to having it imposed upon him. In fact about 50 customers did object, asking for their statements to be supplied with full narrative as usual. It was felt that these requests must be complied with as the branch was unwilling to displease old and valued clients. However, this concession meant that at the beginning of each month special statements had to be prepared for all these customers and, as at King George's Place, this led to a great deal of additional work. At the same time managers and cashiers had the delicate public relations task of trying to persuade conservatively minded customers to accept the new system.

Different members of the Ringer Square management team were quite prepared to talk about the effect of the computer on their jobs. The manager, for example, thought that the computer had made his life more difficult. He missed not being able to examine customers' accounts in detail and during the change he had had to pacify a number of indignant customers. However, he fully accepted automation and appreciated that from the bank's point of view, if not the manager's, it was beneficial. In contrast, the assistant manager did not think his job had been greatly affected. He found it necessary to make a mental adjustment when examining computer printouts instead of customers' ledger cards, but he was quite prepared to accustom himself to the demands of an automated system. The accountant also felt that his job was not basically altered, although he found himself getting the backwash of problems arising out of the new methods. He described three of these; the difficulty of tracing errors in the tape rejected by the computer, the fact that it was now harder to check the state of a customer's account, and the need to time and organise events to meet the demands of the computer—the branch no longer imposed its own deadline, this being done by the computer.

Some of the senior management at Ringer Square were worried at the effect of punch work on the girl machinists. The branch was situated next to the offices of a large retail company and there were rumours that this company had to provide pop music and songs in order to keep its punch operators awake. The supervisor of machinists was not too disturbed, however, and thought the girls liked their new machines. She herself approved of 'the way the work is all done for you—you just punch holes'. She pointed out that with the old system it had been necessary to have one girl for ledgers and one for statements, and this doubled the work. The supervisor's greatest problem was now the tracing of queries. She found these hard to identify without the detailed account information of the pre-computer era.

Generally, management opinion in this branch was that the computer had brought benefits to the bank as a whole, but not to Ringer Square Branch or the individuals employed there. Management found the new system more inflexible than the old; everything was now geared to time schedules and a fixed pattern of work. Both staff and customer adjustment to the new procedures had presented problems, but customer relationships had presented the branch with its biggest headache.

Ringer Square did not reduce its staff numbers as a result of the computer, although some experienced staff were promoted and replaced by less experienced. Like other branches, business had increased and this took care of any clerks who might otherwise have been underemployed. One new post was created, that of computer controller. The holder of this acted as a communication link between the branch and the computer centre, and he was responsible for preparing account amendments for the computer and checking print outs.

It can be seen that from a managerial point of view the principal problems were customer relations, communication difficulties and the need to learn and adjust to a new and unfamiliar system of work. Any of these problems could arise with a modern computer installation and, because work would be transferred so much more quickly, the time for learning and adjustment would be drastically reduced.

Attitudes to work after the change

Because of the bank's slowness in getting its new system into operation, there was a gap of eighteen months between the pre-change and post-change interviews in the St. John's Avenue and Paul Street branches, and a gap of nearly three years between interviews in the other branches. In all, five years elapsed between the first and last interviews. By this time over half our original sample had either left the bank or been transferred to other branches.

We found that in 1964 and 1965 bank clerks were essentially the same kinds of people they had been in 1960. All the men still had grammar school educations and G.C.E., but this time a larger percentage of the women had been to grammar school and 90%, not 50%, now had G.C.E. This surprised us because we had anticipated that the increased routinisation of work would mean that the bank might wish to recruit a lower grade of female staff. A G.C.E. seemed an unnecessarily high qualification for the simple job of punch operation. However, it seemed that the bank was engaging more highly qualified staff than before, as, in addition to G.C.E., 53% of the women had also had a period of further full-time education after leaving school. Despite these higher qualifications the general feeling was still that women couldn't get very far in banking. One shorthand typist told us 'I wouldn't stay in the bank. When I've got as far as I can in the branch I shall leave and try for something I really want'.

Whereas in the pre-change study 76% of the clerks had been trade union members, we found this percentage had now gone down slightly to 72%. This meant our early hypothesis—that office automation would stimulate union membership—was not supported in banking. In fact, when we asked clerks what part N.U.B.E. and the Staff Association had played in the introduction of the computer, there was almost unanimous astonishment that we thought there was anything either of these organisations could have done. Their members did not see supervision of the introduction of technical change as a trade union function.

There were no significant overall changes in what the clerks said they liked most and least about their jobs, but there was some indication that job satisfaction had altered. Our pre-change indices of job satisfaction had suggested that men were more work identified than women, with men in the E.D.P. branches showing higher satisfaction than North Office men. Women were less happy with the task they were doing, but seemed to like other aspects of the work situation—social relationships, for example. We had anticipated that the effects of the computer would not be greatly felt by the men and so their attitudes would not alter, but that the morale of women clerks might deteriorate with the impact on them of the increased rationalisation of work. In fact, the reverse occurred. The attitudes of the women remained essentially the same or even improved—although the machinists were an exception here—while the men in the E.D.P. branches now appeared less contented than the men in North Office. This was most unexpected and we were anxious to find out why this switch in attitude had occurred. Had the computer caused considerable and unanticipated changes to occur in the jobs of the male clerks, or was their reduced job satisfaction due more to a fear of what was going to happen in the future? Why were the women so much less concerned than the men—did their lack of involvement in work mean that change was unimportant to them?

The table below shows how the percentages of clerks saying that they liked the 'interest' of their work, and that they believed their promotion prospects to be good, altered over time. In the post-change analysis we could not use two of the pre-change indices of morale—the clerks' willingness to become a bank clerk again, or his desire for his son to become a bank clerk. This was because the numbers of new clerks in the post-change interview group were too small.

It can be seen that, whereas more women and more North Office men now said 'its interest' when asked what they liked most about their job, fewer E.D.P. branch men said this. Also, whereas there was little change in the North Office men's estimate of their promotion opportunities, E.D.P. branch office men, who had previously thought their promotion chances good, were now *far* more pessimistic. It is clear that our original hypothesis that the computer would be seen as providing men with more promotion opportunities, was incorrect. On the contrary it appeared that men with experience of the new automated system thought the development of E.D.P. limited their opportunities for advancement. This could be because they saw an extension of computer usage leading to an amalgamation of branches and a reduction in

TABLE 11 *Indices of job satisfaction*

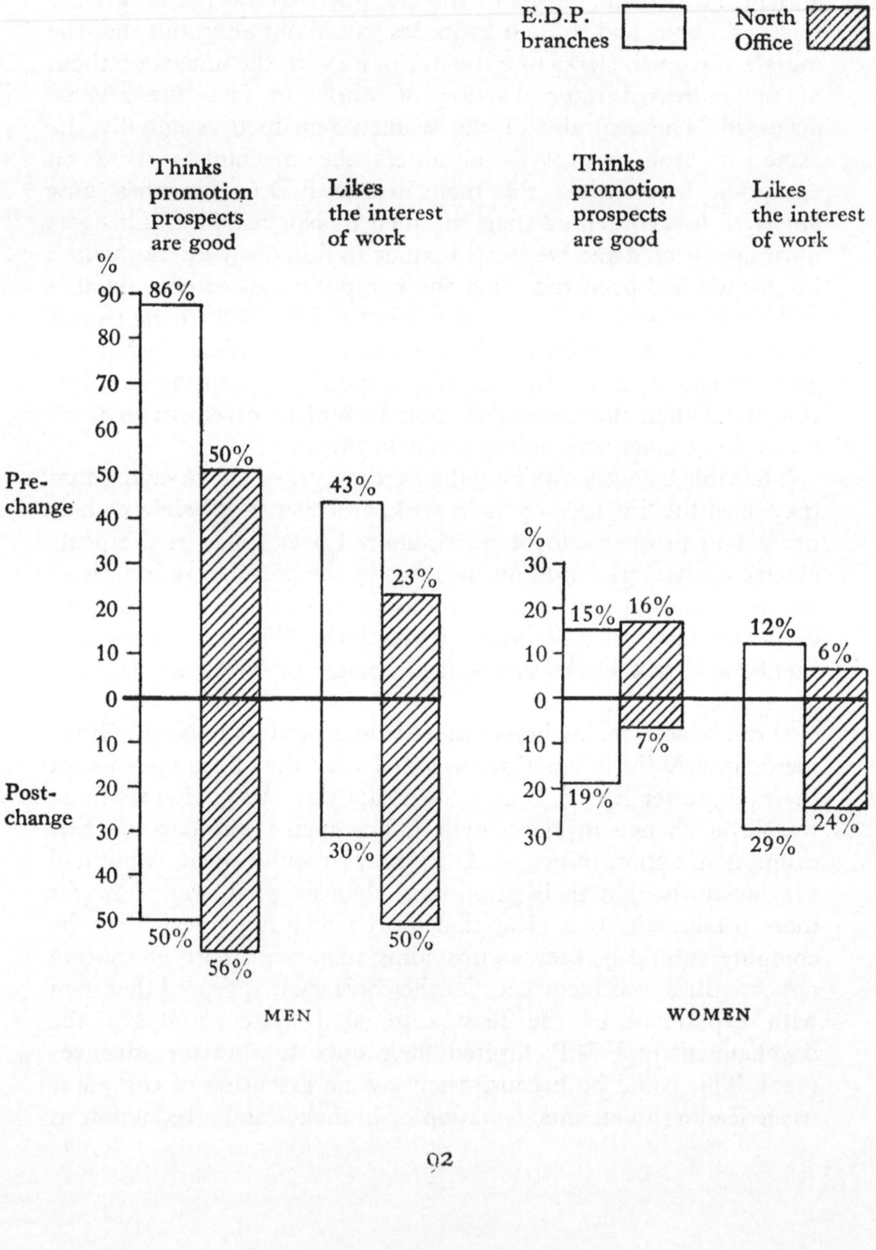

the number of senior posts available, or because they were generally uncertain how computers would affect banking in the future.

How the clerks thought the computer had affected the branch and the bank

E.D.P. BRANCHES

The clerks who had experienced both pre- and post-change conditions in the E.D.P. branches (39 in all) were asked to describe how the computer had affected the branch and, in particular, what the machine's overall advantages and disadvantages were. There was a wide diversity of opinion on the ways in which branch work had changed, clerks tending to speak subjectively in terms of how they personally viewed office automation. For example, one supervisor who saw little good in the computer suggested:

> 'The branch is not as efficient as it was; the time of finishing varies and there is more late work; there are more routine jobs now; there is more paper now, we have screeds and screeds of it; the new system is more messy than the old and the customers don't like it.'

Another had a more favourable point of view:

> 'The computer has increased the accuracy of posting and also the neatness of work. It should take some of the drudgery away.'

The numbers of clerks making favourable or unfavourable comments is shown in Table 12.

The consensus of opinion on the machine's advantages was that work had now become easier or faster (39%) and that the branch now had less work to handle (35%).

Clerks saying work had become easier or faster often appreciated the potential of the computer. One thought 'when the new system is accepted and understood it will be more accurate than the old', another 'the computer will make the balance period easier and generally ease the compilation of returns'. Considering that a large part of branch work was now being carried out by the computer centre, it is surprising that more staff did not mention a reduction of work. However, we have already referred to the fact that the bank's business was continually increasing. Also, at the time of the post-change interviews many clerks were still struggling with a new and imperfect system which had not yet had the bugs taken out of it. This was especially true of King George's Place where comments such as 'we're working much later at night', 'we're working under much greater pressure', were common.

The most prevalent view on the computer's disadvantages (given by 35%) was that the new system was leading to a deterioration in customer relations and a reduction in customer service. It seemed that the clerks' anxiety about this possible consequence of the new system had grown considerably since the pre-change interviews. The Ringer Square staff, with their old family clients, felt most strongly about this, and a control clerk told us 'the old lady customers don't like it one bit; we've lost the personal touch because customers can't get statements when they want them.' While over a third of the clerks interviewed thought that branch work had become easier, a quarter believed that aspects of this had become more difficult. Nearly half the staff interviewed at King George's Place, where the clerks were still struggling with problems arising out of the technical defects of the computer, said this. A number of clerks also suggested that branch work was now too dependent on the computer.

It can be seen that there were no very dramatic changes in the E.D.P. branches as a result of the computer's introduction. The machine's principal overall advantages were a speeding up of work procedures and a reduction in the number of branch account processes through the transfer of ledger and statement posting to the computer centre. These advantages were somewhat offset by the increased difficulty of tracing errors and by the fact that detailed account information was no longer kept at the branches. However, the majority of staff believed the computer had been a success from a bank point of view and that electronic data processing was the best way of coping with rapidly increasing business demands.

NORTH OFFICE

Let us now see if there were any differences in the way that the non-involved clerks in North Office viewed the impact of the computer. North Office clerks were asked how the computer had affected the bank, whereas branch clerks had been asked how the computer had affected the branch. Both groups appeared to have the same things in mind in their answers, however, and so it seems valid to compare their replies. The table below does this.

TABLE 12 *The clerks' views of the computer's effect on the bank/branch**

	Entirely favourable	Neutral or ambivalent	Entirely unfavourable	Don't know	TOTAL
North Office	4	6	11	13	34
E.D.P. Branches	10	20	8	1	39

* New clerks were not asked this question.

We found that those North Office clerks who were able to express an opinion viewed the computer's effects more unfavourably than E.D.P. branch clerks. North Office men were more critical than North Office women, over a half of whom said 'don't know' in answer to the question—an indication perhaps of the woman clerk's lack of interest in the changes that were taking place. All except one of the clerks making critical comments referred to the deterioration in customer service; 'the customers don't like the new system'; 'the service to the customer is not as good now'. North Office clerks even more than E.D.P. branch clerks thus appeared to think that the new computer system was damaging the bank's image.

As we have seen, customer problems were frequently mentioned in the E.D.P. branches, but only when clerks were asked specifically to state the computer's disadvantages.

On the computer's advantages for the bank, North Office opinion was similar to that in the E.D.P. branches. Over half referred to faster work or more accurate work, the response we had received most frequently from the other group. One difference was that 17% of North Office staff gave the machine's potential for cutting down staff as an advantage from the bank's point of view. No-one suggested this in the E.D.P. branches and it would seem that North Office staff were more conscious of the possibility of automation reducing employment opportunities for bank clerks.

The disadvantages of the computer for the bank were again that the new system had led to a deterioration in customer relations. 'There will be queues of irritated customers', said one young man; 'it's taken away the personal touch,' said another. Men were more likely to refer to customer problems than women; the most frequent feminine response was still an uninterested 'I don't know'.

It can be seen that although North Office staff were not really affected by the computer this did not make them more favourable in their attitudes towards it. On the contrary they viewed its consequences more critically and apprehensively than E.D.P. branch clerks, and particularly its consequence for the customer. Also, whereas E.D.P. branch clerks had seemed confident that banking was still a secure occupation, North Office clerks saw the computer as reducing staff. It could be argued that this increased awareness of the likelihood of staff saving was due merely to the passage of time, and to a greater general knowledge of office automation and its consequences. However, we had interviewed in Ringer Square immediately before visiting North Office and had not found this reaction there. A possible explanation of this difference in attitude is that E.D.P. branch staff saw for themselves that the computer had so far caused little staff saving and they expected this state of affairs to continue. They believed business would expand faster than automation could eliminate jobs.

The most striking point emerging so far from this comparison of attitudes to change in affected and non-affected branches is that a group spared the impact of change will not necessarily view it more favourably than those directly involved. North Office clerks, perhaps to emphasise their own good fortune in having escaped the upheaval, seemed to have a very pessimistic view of what was going on in the E.D.P. branches. A subsidiary point is that women who are in a situation which produces work apathy and little opportunity for advancement will not care very much about events outside their own departments. Their energies and thoughts are transferred to social relationships and leisure activities.

How the clerks thought the computer had affected themselves

E.D.P. BRANCHES

Let us now see how the clerks thought that they, as individuals, had been affected by the computer, starting by looking at the different experiences and perceptions of men and women.

When clerks were asked to analyse how the computer had altered different aspects of their job, the following pattern emerged.

TABLE 13 *The effect of computer on different job aspects**

(The question asked was 'How has the computer affected the following aspects of your job—The degree of accuracy in your work—The amount of work you have to do?' etc.).

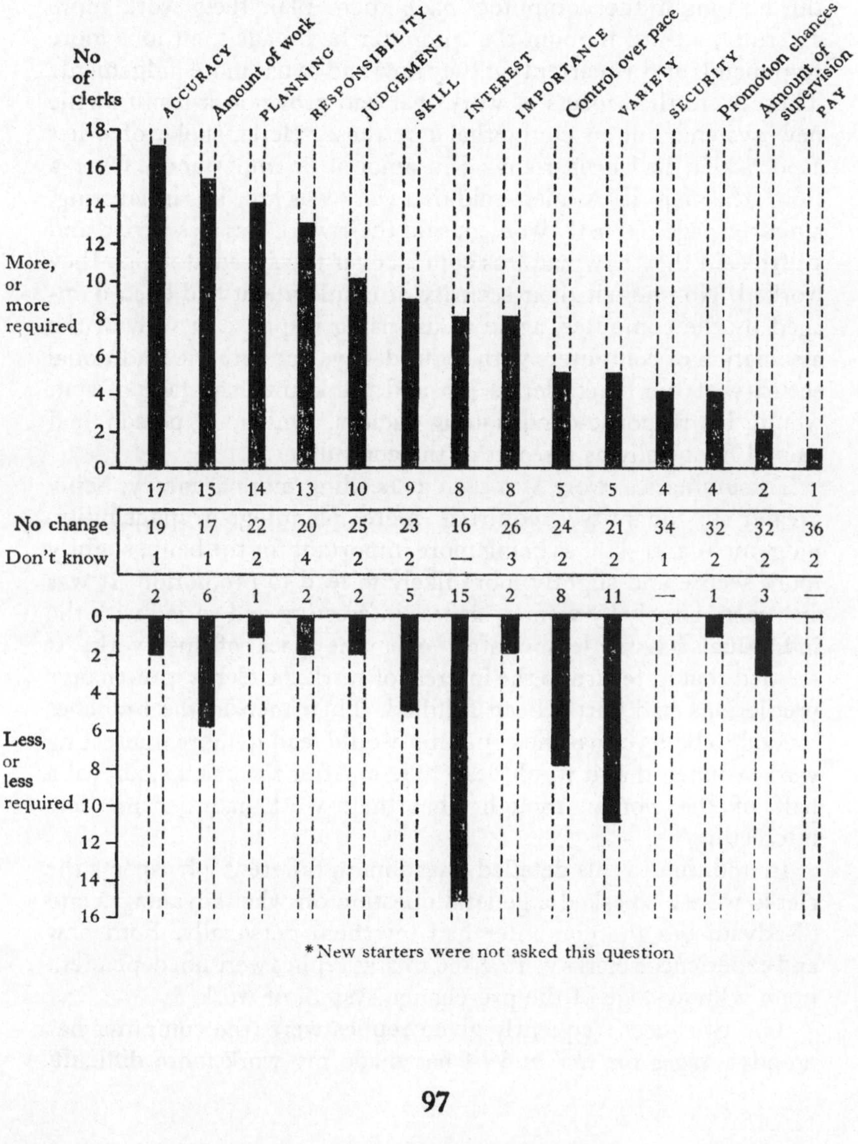

	ACCURACY	Amount of work	PLANNING	RESPONSIBILITY	JUDGEMENT	SKILL	INTEREST	IMPORTANCE	Control over pace	VARIETY	SECURITY	Promotion chances	Amount of supervision	PAY
More, or more required	17	15	14	13	10	9	8	8	5	5	4	4	2	1
No change	19	17	22	20	25	23	16	26	24	21	34	32	32	36
Don't know	1	1	2	4	2	2	—	3	2	2	1	2	2	2
Less, or less required	2	6	1	2	2	5	15	2	8	11	—	1	3	—

*New starters were not asked this question

Table 13 shows that nearly half the clerks believed the computer had increased the need for accuracy in their jobs, and this response was distributed evenly between men and women. Over a third believed the computer had increased the amount of work they personally had to do, an interesting response in view of the fact that computers are usually thought of as reducing work. Over a third thought the computer made them plan their work more carefully, a third thought the computer had made their jobs more responsible and a quarter that they now had to use more judgement. These were the aspects of work that had grown as a result of the new system, but to counterbalance these, clerks spoke of other aspects that had been reduced in amount or importance. Over a third, the majority women, said that work was now less interesting, a quarter, again mostly women, said their work was less varied and a fifth said they now had less control over the speed at which they worked. No one felt their security of employment had been damaged by the computer, again a surprising response in view of the association of computers with redundancy, but here the traditional security of the bank clerk's job and the known shortage of staff would be response-conditioning factors. Only one person had gained financially as a result of the computer.

To summarise, work was seen as needing more accuracy, being greater in quantity, requiring more planning, responsibility, judgement and skill; as being more important for the bank, slightly more secure and slightly more likely to lead to promotion. It was also seen, largely by women, as less interesting and varied with the individual having less control over the pace of his work. It seemed that in regard to the interest of work the clerks' pre-change predictions had partly been fulfilled. Then men in the branches tended to believe that the computer would lead to more interesting work, women that it would lead to less. After the change, almost a half of the women thought that their work had become less interesting.

In addition to this detailed question on different job aspects the clerks were also asked a general question on what advantages and disadvantages the computer had for them personally. Both new and experienced clerks were asked this, as replies were not dependent upon a knowledge of the pre-change system of work.

The two most frequently given replies were 'the computer has no advantages for me' and 'it has made my work more difficult'

TABLE 14 *Advantages and disadvantages of the computer for the individual clerk*

	Advantages				Disadvantages*		
	M.	F.	All		M.	F.	All
Makes work easier in some respect	4	7	11	Makes work more difficult or boring	8	14	22
Reduces the amount of work	1	3	4	Working conditions have deteriorated	3	3	6
Miscellaneous replies	2	1	3	We have difficulties with customers	4	—	4
No advantages	8	12	20	No disadvantages	8	9	17
Don't know, no answer	5	8	13	Don't know, no answer	—	6	6
	20	31	51		23	32	55

*4 clerks gave more than one answer to this question.

('or more boring'), boredom being almost exclusively a feminine response. This disgruntlement was to some extent compensated for by a third of the staff saying the computer had no disadvantages for them, and a fifth that it had made their work easier. But on the whole, the computer was seen as bringing individuals more disadvantages than advantages.

A complaint made by both men and women was that the computer had made work more impersonal. With the old system, clerks had taken an interest in the customers whose accounts they were processing, knew their names and how their financial affairs were progressing. These very real and identifiable clients had now disappeared, for the removal of ledgers and narrative statements meant that a customer's account could no longer be seen as an interesting series of transactions; it had become a balance figure attached to an account number.

Besides this increase in the impersonality of work, girls complained that there were now more rather than fewer routine jobs. The supervisor of girls at King George's Place summed up the post-change situation; 'there is now less interest for the majority and more routine jobs. We have an awful lot of sorting and filing to do. The computer won't sort the cheques for us'. It can be seen that in the bank the computer had by no means eliminated routine jobs.

So far we have been considering only the computer's impact on the two sexes without taking into account occupational groupings. But it will be remembered that post-change observations in the E.D.P. branches indicated that the work of both machinists and control clerks had been considerably altered by the new system. We were, therefore, particularly anxious to obtain the views of these two groups.

Comments on how the 14 aspects of their jobs had changed (Table 13) showed that machinists were far more likely than other clerks to say that work had become less interesting. In fact, all of them made this comment. One machinist told us 'the punch machines are not interesting at all—it's just feeding the computer, on other jobs you see the results of what you're doing, we don't,' another 'the computer has taken away a lot of the interest. Before you knew the accounts of your own customers, now the computer deals with them and, because there is no longer any ledger posting, you don't know what is happening'. Similarly, in answer to the question 'what do you like least about your job?' all the women mentioning routine or monotony were machinists (this had also been the case before the change). Two-thirds of the girls operating machines wanted to change their jobs and all of these expressed a desire to do something more interesting, for example, 'I'd like to go on the counter, you're just a number on the machines.' The most favoured job with these girls was cashier, although shorthand-typing was also mentioned and one girl was prepared to do anything providing it was more interesting than her present job. Machinists were also more likely than other clerks to say their work now lacked variety. One told us, 'I don't feel I'm doing anything worthwhile on the job I'm on. I might as well be in a factory as sitting here machining cheques'.

Machinists were almost the only women saying that the computer had led to some of the desirable elements in their jobs being taken away. They referred to less variety, skill, responsibility etc. The other women clerks mostly found that there had been no major change in their jobs, although a third found that they now had more work, needed more accuracy and had to do more work planning. A manager's secretary pointed out 'we have to be particularly careful with new accounts. One slip up and the whole think goes wrong. Everyone finds it far more important to be accurate with numbers instead of names'. A girl cashier complained 'if we have a dicey customer we need to check up, yet we don't have a full print out of accounts every day'.

However, it seemed that the greatest impact of the computer had been felt by the girls operating the new punch machines and, for this group, the new system brought nothing but disadvantages—increased routine, monotony and boredom. It will be recalled that these girls had grammar school education and had passed their G.C.E. examinations. It seemed that the bank needed to revise its recruitment policy and take on girls who could tolerate this new kind of work.

In the predominantly male groups, apart from branch management, the control clerks were the occupational group whose work was most affected by the computer. Only three of these in our sample had experienced both old and new systems of work, but they were reasonably unanimous in their replies. Like the girl machinists, all of them found that work had become less interesting and varied.

One claimed 'the computer has destroyed the interest of our work; we might as well work in a factory putting tops on bottles'. Another said his job had become 'a hundred times more boring, previously I was in charge of three girls. Now I don't even see the work; it is all put on tape and this is all we see'.

These young men, like the girl punch operators, seemed to be suffering from the de-personalisation of work, and from the removal of a pleasant work relationship with a girl assistant. Two commented that they were now having to do work previously completed by the girls. Also, like the girls, they found that the new system led to a reduction rather than an increase in the different elements set out on Table 13.

Apart from the control clerks, the male clerks seemed to have few criticisms of work changes. There were references to the need for more accuracy and work planning and to the greater responsibility they were required to show, but these were not seen as drawbacks. In view of the slight impact of the computer on men, it is surprising that our earlier assessment of job satisfaction suggested that the morale of E.D.P. branch men had deteriorated since the introduction of the new system. While there is no direct evidence that this was due to experience of office automation—the fact that it occurred in the branches, but not in North Office would suggest this very strongly.

To sum up, the main impact of the computer was felt by certain groups, notably management, girl machinists, control clerks, and customers and all these felt they had lost out with the new system. Managers thought their jobs had become more difficult; they had lost their autonomy, they no longer had immediate access to detailed account information and they had the tricky public relations task of persuading customers to accept the new system. Girl machinists, who had entered banking believing it to be an attractive and interesting job, now found that the computer had made their work more boring and taken away the personal element. Customers who had previously been real people using their money in interesting ways, were now reduced to account numbers and balance totals. These girls also commented that their jobs now resembled machine operators on a factory shop floor. Control clerks, although less affected than the girls, also found their jobs less interesting and missed the intimate knowledge of customers' affairs they had previously known. For these two groups it seemed that work had been routinised and de-personalised by the computer and they did not like this.[3] Although we do not have the views of any customers, the clerks were convinced that they resented the loss of narrative on statements and objected to the new requirement that statements must be ordered in advance.

NORTH OFFICE

North Office clerks were also given the fourteen different job aspects referred to in the previous section, but instead of being asked how their own jobs had changed in each respect, they were asked how they thought the jobs of clerks in the E.D.P. branches had changed as a result of automation. The table below shows their answers set out beside the answers of the branch clerks.

When this table is analysed it is apparent that North Office staff

[3] This de-personalisation of work seems to be a consequence of advancing technology in other situations. Many hospital operating theatre staff are now primarily technicians controlling complicated electronic equipment. Nurses, who like bank clerks, have been brought up in a tradition of personal service, are often reluctant to do theatre work for this reason.

TABLE 15 *Comparison of opinion on how the computer had affected different job aspects*

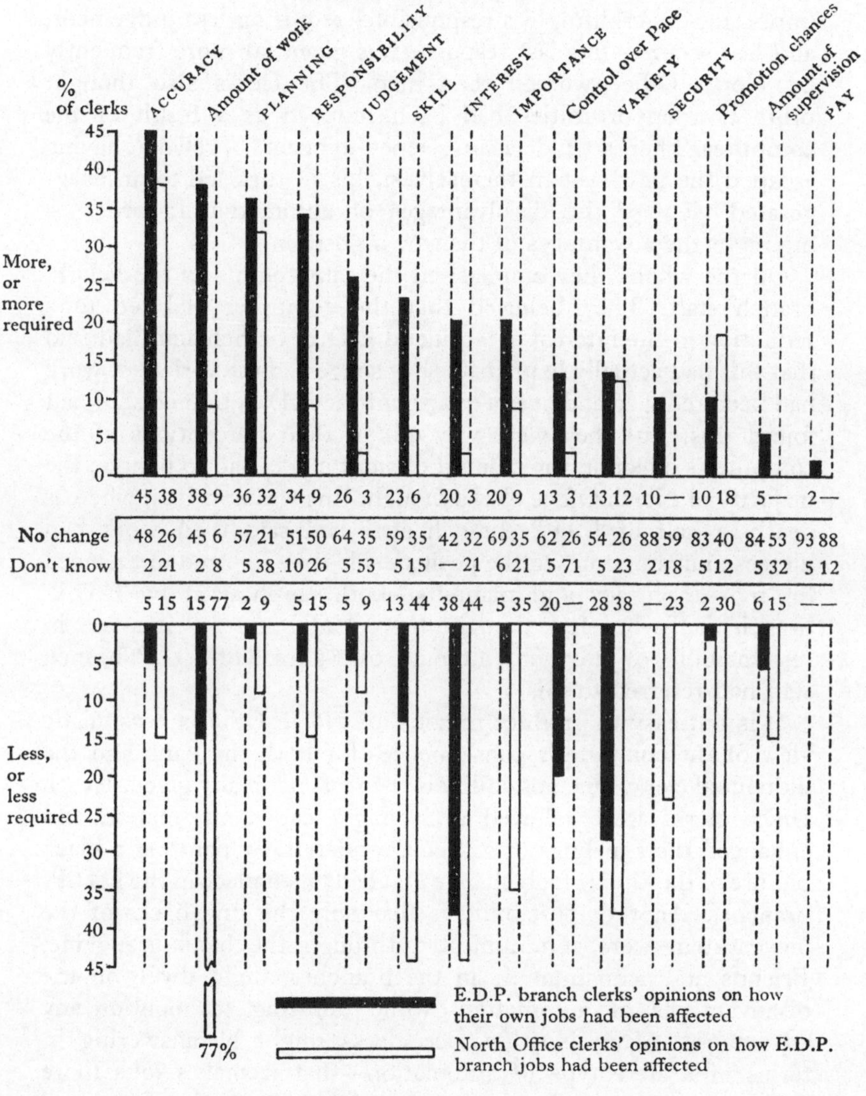

saw a black picture of how the jobs of the E.D.P. clerks had changed, a much blacker picture, in fact, than the clerks who had experienced the changes. They were significantly more likely than E.D.P. branch staff to say the computer had made work less important, less skilful, less responsible, requiring less judgement, and less secure—this last response was given far more frequently by North Office women than men. The clerks also thought promotion opportunities had been reduced as a result of the computer. Their attitude again seemed to be one of relief at having escaped the new system themselves; this in turn led to an exaggerated view of the disadvantages of automation in order to highlight the advantages of their own position.

On the whole, they emphasised the same points as the E.D.P. branch staff. They believed that the computer had led to a reduction in the interest of work and E.D.P. branch staff had said that this had actually happened; they suggested the variety of work had been reduced and over a quarter of the E.D.P. branch staff had found this. But they were way out in their perceptions of the computer's effect on the amount of work clerks had to handle; the majority of North Office clerks thought this was now less, whereas nearly a half of E.D.P. branch staff had said their work had increased in amount. On the positive side they referred to the need for more accuracy and planning—both emphasised by E.D.P. branch staff—but few thought there had been any increase in responsibility or judgment, although over a third of E.D.P. branch staff had referred to this.

It is not easy to produce reasons for North Office's pessimistic view of the computer's consequences for both the bank and the individual clerk. One possibility is the reason already given. North Office clerks were relieved at having escaped automation and enhanced their feelings of self congratulation by painting a black picture of the kinds of job change that had taken place in the E.D.P. branches. Another possibility is that only the drawbacks of the new system were communicated through the bank grapevine. Friends and acquaintances in the branches might dwell on the disadvantages of automation, while omitting to mention any advantages. Again North Office clerks might be answering in terms of a stereotype of automation—that it makes jobs more boring, causes redundancy etc.—but doing this without any real information concerning what had taken place. Or, they might have been generalising from their own slight experience of automation. The first two suggestions seem the most plausible. With the exception of their remarks on the amount of work saved by the computer, the North Office clerks emphasised the same features of automation as the branch clerks, although always more negatively. This could mean either that they were receiving some communication from the branches, but that the information sent out was of a pessimistic kind, or that the North Office clerks were selective in what they received, accepting critical comment of the new system and ignoring approval.

Against this theory is the fact that informal communication between the E.D.P. branches and North Office seemed poor. When we asked North Office clerks how they had heard about the computer and the new system of work, only four of the 47 clerks gave friends in other branches as their source of information, although a number said vaguely that they had picked up the information, or that they had heard others talking. Also, as we shall see in the next chapter, North Office clerks knew virtually nothing about the problems branches had experienced when transferring to the computer, indicating that there was very little communication in this area. One girl had heard there had been late working at Paul Street, another said she knew that accounts of one branch had been in a mess. A man referred to the computer breaking down. Apart from this there was no comment.

The sparseness of informal communication was certainly due to the physical separation of the branches from each other and the lack of contact between a scattered staff. It seems probable that North Office clerks had gleaned some information about the job changes that had taken place in the branches from gossip and rumour. Using this they arrived at a conception of automation and its consequences which reflected their attitudes to a system they didn't understand, knew little about and didn't welcome.

Clerical work as a career

We were not only interested in finding out how the clerks thought office automation would affect the bank and bank clerks, but also in their opinion on the wider question as to how it would affect clerical work as a career. We therefore asked them: 'Has the

TABLE 16 *How the clerks thought computers had affected clerical work as a career (E.D.P. branches)*

	For the worse	Made no difference	For the better	Don't know, no answer	TOTAL
For girls	24	6	7	2	39
For boys	9	19	6	5	39

introduction of computers into offices affected clerical work as a career for (a) boys and (b) girls?'

Nearly two-thirds of the clerks in the E.D.P. branches thought computers had damaged the clerical career prospects of girls, and this answer was given nearly as frequently by men as by women. On the other hand the majority believed that the career prospects of boys had remained unchanged. Thus women, rather than men, were seen as the victims of office automation. Clerks suggested that office work would become less interesting and varied for girls and that there would be fewer opportunities for them to get on and make careers for themselves. Commercial courses for girls were seen as becoming redundant, for the computer would take over these functions. One young man suggested 'in time we may need similar girls to those who work in factories, if the trend towards machines continues.' Another, 'their work is now on a level with factory work.' A less threatened future was forecast for boys, 'they won't be affected in any way' said one young man firmly. Others were not so certain about the long term effects of automation. 'There won't be any effect on boys for some time, but in the future fewer may be needed, and their jobs may become routine.' This masculine complacency does not fit in with the male clerks' earlier statements that their own promotion prospects had become less favourable since the computer. It can only be suggested that they did not think of themselves when answering this question, but spoke in general terms. However, the answers do show a change in attitude as a result of direct experience of automation. In the pre-change interviews nearly half the clerks in the branches had thought computers would have favourable consequences and only one-fifth had thought the results would be adverse. Now those speaking favourably of clerical prospects were reduced to a sixth.

Table 17 shows how North Office clerks thought computers had affected clerical work as a career.

TABLE 17 *How the clerks thought computers had affected clerical work as a career (North Office)*

	For the worse	Made no difference	For the better	Don't know, no answer	TOTAL
For girls	14	8	4	8	34
For boys	7	16	5	6	34

The distribution of answers is essentially the same as in the E.D.P. branches—a worsening of career prospects for girls, but little difference for men. Many clerks again depicted a gloomy future for girls. 'There will be more routine jobs in the future; the bank will no longer look for clever girls', 'I wouldn't advise girls to come into the bank; things have gone worse for them.' Some of the young men who saw no change in the career prospects of girls felt that the situation was unsatisfactory already. 'Girls get a raw deal now; the computer won't change it.' Others seemed to think girls would accept anything. 'We already have girls with G.C.E. "O" levels who are willing to do monotonous work, so why should the situation change?'

It was the women, not the men, who saw automation leading to a deterioration of clerical work as a career for boys. One girl told us 'yes, it will be worse for them. They've got to do it all their lives.' The men themselves had few fears in this respect. 'More supervisory jobs may be available in the future. The bank may take fewer men, but those who come will have better opportunities.' 'The good will be able to get better jobs than before, although the worse will get less good ones than in pre-computer days because a chunk of work has now gone'.

Again we come up against this enigma. Girl clerks, most of whom were uninterested in what was happening in the bank, were well aware of a possible threat to female careers, although they seemed relatively unconcerned about this. However, men who were showing signs of anxiety about office automation and their own promotion prospects still believed that masculine opportunities in clerical work would not generally be reduced.

Conclusions

What conclusions of relevance to managers can be drawn from this analysis of branch opinion? The first is the rather obvious one that men, with their vested career interests, are likely to be more anxious about technical change than women. Young girls working on routine tasks—providing these are not adversely affected by a new system of work, and this is important—are not sufficiently work orientated to care much what happens in the employment situation. Management then needs to pay particular attention to identifying its male employees with any changes that are taking place. It may also need to consider its older women staff. These are likely to be more career minded than the young and more anxious about change. Although there were few mature women in our sample the proportion of this group in the working population is increasing and likely to go on doing so.

A second point is that clerks performing routine accountancy tasks, or any routine work for that matter, appear to get considerable satisfaction from being able to identify themselves with the clients whose affairs they are handling. They learn the names of their customers; observe their financial transactions and the solvency of their accounts with much interest. Managers must appreciate the undesirability of de-personalising work to such an extent that real people become anonymous account numbers and clerks have the last bit of interest removed from very boring work. A great deal is now being written about the alienation of the shop-floor worker from his job.[4] Work becomes so divided up and routinised that it provides no satisfaction in itself and is merely a mechanism for earning money. Office automation need not cause this to happen to the white-collar worker if efforts are made to retain or build interest into work. We suggest that job enlargement is as possible in the office as on the shop floor.

Thirdly, attention needs to be paid to the new routine and unpleasant job of key punching—that is punching tape for a computer. This job is both boring and exacting, as it requires speed, accuracy and concentration. It is a solitary job and there is little possibility of social interaction while it is being performed.[5] In fact, it has all the characteristics of the kind of work that social

[4] Blauner (1964) provides two detailed examples of shop floor alienation —textile workers and automobile assembly line operatives.

[5] We shall, however, be showing in Chapter 8 that some girls liked key punching.

psychologists have always described as the most tedious and stressful. Punch operators have been described as 'nervous wrecks who stay home often and keep supplies of tranquillisers and aspirin in their desks at which they are "frozen" as though it were a spot on an assembly line'. (Baker, 1964). It seems managers need to give thought to alleviating some of the disadvantages of this kind of work. An appropriate selection policy would seem the first thing to consider—recruiting girls who can tolerate routine work and not, as Royal Exchange did, looking for girls with educational qualifications such as G.C.E.[6] Again, such things as job rotation, frequent rest breaks and the opportunity for permanent transfer to other work after a known and limited period of time would all help to make punch work more tolerable.

Our analysis provides us with two other warnings for those introducing computers. The reactions of men in the E.D.P. branches show that morale can drop in groups involved in technical change, not so much because their immediate situation is radically altered by the change, but because they read into the change implications that they see as affecting their future career prospects. Also, North Office opinion on what had taken place in the E.D.P. branches indicates that people not involved in a change can take an even blacker view of what is happening than the participants. This suggests that managers introducing change must, in their communication policies, pay as much attention to non-affected as to affected groups if they wish to secure majority approval.

[6] Personality tests could prove useful here. The work of Cooper and Payne, psychologists in the Department of Industrial Administration, University of Aston, Birmingham, suggests that the personality factors of extroversion and neuroticism are related to an individual's ability to tolerate routine tasks. After studying the behaviour of female operatives on a packing line they concluded that the more extraverted an individual is, the less he is able to tolerate routine tasks.

CHAPTER 6

The Royal Exchange Bank—An analysis of attitudes after the change

In our pre-change analysis we tried to discover reasons for some clerks looking favourably on the 'idea' of office automation and others reacting against it. We could find no relationship between factors in the clerk's background—such as social class position or education—and pro- or anti- attitudes to change, and it seemed that one of the most important influences on opinion was how the clerk predicted the change would affect him personally; whether he saw it as helping or hindering his own objectives and interests. For example, girls with G.C.E. were against the idea of a computer because they thought it made their qualifications irrelevant.

In our post-change analysis we can establish whether the same variables influence attitudes to the 'reality' of change and, if so, whether these attitudes take the same direction as in the post-change. Will girls with G.C.E. still be against the computer or will they now be neutral or for it? If attitudes have altered then we must see which groups have changed their opinions and try to establish why they have done so. The views of North Office clerks will help us establish whether actual experience of office automation is the reason for changed attitudes. If this is the case then the post-change views of clerks in the E.D.P. branches will not necessarily be repeated in North Office. However, if similar opinions are found both in the E.D.P. branches and in North Office, then there are two possibilities. Either informal communication between the two groups is good and North Office clerks are identifying with clerks in the E.D.P. branches and adopting their attitudes—but our data have already shown that informal communication was poor—or, more likely, actual experience of office automation is a less important influence on attitudes than a general and growing awareness of the implications of computers for the future career prospects of clerks.

In this chapter we will also examine the influence on attitudes of another variable—the way the change was introduced and carried through by management. Obviously, the way change is interpreted by employees must partly depend on how it is presented to them. Clerks who see office automation as threatening can have their fears reduced if honest reassurance is given by the innovating group. Similarly, clerks who are ambivalent or neutral may become approving or hostile according to whether change is presented and handled well or badly. Introducing change is a great challenge for management because it provides an opportunity for making policies explicit. 'We will do so and so if such a thing happens' becomes 'This has happened or is going to happen, therefore we *are* doing so and so'.

Some factors influencing attitudes

In our pre-change analysis we found that men who thought their promotion chances were good looked favourably on the idea of office automation, whereas men who found their present jobs interesting were neutral, and men who did not particularly like their jobs were indiscriminately either for or against the proposed change. Women reacted rather differently. Young girls who had always worked in the same branch approved of the idea of change, but girls with G.C.E. qualifications considered it a threat and were opposed to it.

In the post-change analysis we compare the pre-change attitudes of individuals to the 'idea' of office automation with their post-change reactions to the 'reality' of the new situation. We have also tried again to identify characteristics affecting attitudes to change. This has been done by testing all our variables—sex, age, education, attitude to work, promotion prospects, etc., against the answers to four questions. These are:

1. In what ways has the branch been affected by the computer?
2. What advantages and disadvantages does the computer have for the branch?
3. What advantages and disadvantages does the computer have for you personally?
4. Has the introduction of computers into offices affected clerical work as a career for (a) girls, and (b) boys?

In this way we can discover if clerks giving the same kind of response had anything else in common with each other.

E.D.P. BRANCHES

Again one of the most striking features of our results is the difference in the attitudes and reactions of men compared with women. With the exception of the machinists, whose problems we discussed in the last chapter, the attitudes of the women after the change were essentially the same as they had been before. 70% of the group who were interviewed twice were quite consistent in their pre- and post-change attitudes. If they believed before the change that the computer would be a bad thing for clerks, they still believed this afterwards. There has been no increase in approval of the computer or hostility towards it. Only six men were interviewed twice in the E.D.P. branches and so pre- and post-change comparisons of this small group are of dubious value. But only two of the six held the same opinions in both interviews. Those who were neutral or favourably disposed to office automation before the change were against afterwards and vice-versa.

Our attempt to identify factors influencing the post-change attitudes of women clerks proved fruitless. After the change girls with G.C.E. were still more likely to be against automation than other girls, but the difference was no longer statistically significant. A likely reason why we drew this blank is that all our questions were directed at ascertaining information about bank clerks in relation to their work situation. But, as we have seen, the girls were not particularly work-orientated and the most important factors determining their attitudes probably lay outside work and in their social situations. If a girl was engaged to be married, or not proposing to stay in the bank much longer, whether a computer came or not would be irrelevant to her. Only the machinists were affected by the new system in a major way, and only this group produced any violent reactions.

The men, however, were in a very different situation. They were committed to banking, the rest of their lives would be spent in Royal Exchange, and the future organisational consequences of the new system were therefore very important to them. With this group we were able to identify a number of characteristics seemingly influencing attitudes to change, and all of these were quite consistent with each other. Men who had been in a branch before it went on to the computer were less likely than others to say its consequences for the branch were desirable. Men who thought the computer had disadvantages for them personally were less likely to say it had advantages for the branch. Those who said they particularly liked the interest of their work were less likely to approve of the computer. Those who thought their promotion chances were good were more conscious of its disadvantages—this last point represented a swing in attitude, as before the change men with good promotion chances were more likely to be in favour of the computer. Men who wanted to change their jobs for promotion were less likely than others to say computers had improved the career prospects of boys, and those who had made plans for their future careers, usually in terms of promotion, were more likely than others to refer to the disadvantages of the computer for the customer. In contrast, men who didn't like anything in particular about their jobs were more likely to be in favour of the computer.[1]

From this statistical jigsaw it is possible to arrive at a picture of the kind of man who, in Royal Exchange bank at least, is more likely to oppose than to welcome innovation. He is probably one of the bank's most esteemed employees, identified with the interests of the bank and fitting in well with the traditional patterns of banking. He likes his job, is very conscious of the bank's aim of maximum customer service, has plans to get on, and believes he can achieve this ambition. The majority of experienced male bank employees would fit this description, and their response to the reality of automation probably accounts for the drop in morale shown in our post-change indices. Although the kind of person likely to welcome change emerges much less clearly from our results, it seems that he is young—men under 21 were very much more in favour of automation than others. He is not yet identified with the bank or particularly interested in his work, but he likes the idea of change and progress. Although unexpected, these reactions seem logical. The young man has little to lose, he is only at the beginning of his career and, as yet, not particularly interested in it. The rather more mature man is in a more vulnerable position. He has given around ten years service to the bank; he probably has

[1] The relationship between men who said they liked the 'interest' of work, that their promotion opportunities were good, and that they wished to change their job for promotion—and an adverse opinion of the computer—is significant only at the 10% level.

acquired domestic responsibilities; he is approaching the age when managerial posts are becoming attainable, and he believes he has a good chance of obtaining one of these. Any alteration in a hitherto almost entirely predictable future introduces uncertainty and therefore is threatening to his objectives. He wants things to stay as they always have been, not to change.

The problem for the bank, then, is not with its young male clerks. They are adaptable and part of the computer age; they will welcome automated banking and grow with it. The real difficulty is in persuading the reliable, experienced, and rather older employee—although he is still under 30—who is identified with the bank and its interests, that automation is not a threat to his future career prospects. It seems that the Royal Exchange bank had not succeeded in doing this although the bank's management had emphasised to staff on many occasions that automation should improve, not damage, career prospects.

Although men and women had different attitudes and reactions on most things, there were a few common factors influencing their attitudes in the same direction. Youth was one of these. All bank clerks who were 21 years of age or less were more likely to see the effect of the computer as favourable for the branch. Similarly, all clerks who were not particularly interested in their work and who said there was nothing special they liked about their jobs were more likely to view the new system of work favourably. In contrast, clerks saying they liked the 'interest' of their own jobs were very conscious of the disadvantages of the computer for the customer. This group identified strongly with the bank and its objectives.

One further point of particular note emerged from our results. Clerks who had joined either the Bank Staff Association or the National Union of Bank Employees were very much more likely than other clerks to see the effects of the computer as unfavourable. There could be two reasons for this relationship between union membership and an adverse attitude to office automation. First, the fact of joining a union suggests interest in and identification with banking. The young or apathetic clerk would not bother to do so and it is, of course, this last group which approved of change. Second, both the Bank Staff Association and the National Union of Bank Employees, especially the latter, were concerned about the effects of office automation on their members. This anxiety could easily have been passed on at branch meetings and in union publications.

This analysis of factors affecting attitudes to change leads us to an interesting conclusion. In banking, and this may be true of other office situations, resistance to change will not come from the apathetic individual who does not like his job, is not involved enough with it to join a union, and has not got the company's interests at heart. On the contrary this kind of person may be for change because he will not see it as threatening his interests. Work means little to him, therefore what happens at work is to a large extent irrelevant. From a management point of view this is a disturbing result, for the people who will see change as undesirable and threatening are likely to be management's most approved of employees, those clerks who are highly identified with company goals and anticipate a worthwhile future for themselves in a job they enjoy. In occupations like banking, where there is a tendency to recruit conservatively minded staff, who like security and make conscientious and responsible, but not dynamic, employees, there may be difficulty in getting major change introduced without stress and anxiety. Organisations which recruit young extroverted individuals, who put their own interests first and are less company minded, may find the introduction of change a much easier matter.

NORTH OFFICE

We were particularly anxious to find out whether the same variables affected attitudes to change in a branch that had not been greatly altered by the computer. If this proved to be the case it would support the view that it is not just the experience of change and the nature of its impact that influences reactions. A knowledge of what is taking place and an evaluation of the implications of this for the future, may be equally important.

In the E.D.P. branches we had found little consistency in the pre- and post-change attitudes of the male clerks who were interviewed twice and we also found this in North Office. Only two out of ten gave the same kind of comment in both interviews; one of these was against the computer on both occasions, the other was neutral towards it. As almost all the women replied 'I don't know' in answer to the question 'What effect has the computer had on the bank?' no before and after comparisons could be made with this group. It seems reasonable to conclude, however, that in both the E.D.P. branches and North Office the male clerks' perceptions of how the computer was going to affect both themselves and the bank changed over time and caused pre-change attitudes to be modified or reversed. In the E.D.P. branches actual experience of the change caused this modification; in North Office it came about through increased understanding of the implications of office automation.

Both groups and individuals altered their opinions about office automation and in North Office we found similar post-change reactions to those we had identified in the E.D.P. branches, particularly in regard to men. In the E.D.P. branches, after the change, it had proved impossible to identify any factors other than occupation influencing the attitudes of the women clerks towards the new system. In North Office it proved equally difficult; again the strongest impression was one of feminine apathy leading to a lack of interest in what was taking place in the branches. The G.C.E. girls who, in the pre-change study, had been particularly antagonistic towards the idea of a computer, in the post-change study had modified their critical approach and were as ignorant about the change as the less academically qualified girls. In North Office the occupation of women clerks did not seem to influence their attitudes. Whereas E.D.P. branch machinists had reacted strongly against the computer, finding it a mechanism for making their work more monotonous, North Office machinists were no different in their attitudes from girls doing other jobs.

North Office men, like the branch men, were in a happier situation than the women. They were far more likely to say that their promotion prospects were good, to have career plans, to want to change their jobs for promotion and to say they found their work interesting. Because of their vested interest in banking as a career they had made some attempt to obtain information on the changes that had taken place in the branches and they were prepared to comment on these. We found that the same kind of man was reacting in a similar way to the new system in both the branches and North Office, even though North Office attitudes were based on hearsay, not experience. For example, North Office control clerks were significantly more likely to say that the computer's effects were unfavourable for the bank than other male clerks—it will be remembered that E.D.P. branch control clerks thought their jobs had been adversely affected by the computer. Similarly, men in the 25-30-year-old group were more likely to be critical of the computer and its consequences than younger or older men, and this group were also more likely to see the computer as damaging bank and customer relationships. We had found the 20 to 30 age group equally critical in the branches. Men with good promotion prospects, hence strongly identified with the bank, also saw the probability of adverse customer reaction as the computer's principal disadvantage. We had observed a similar reaction in the E.D.P. branches and in both cases it represented a switch in attitude, as the pre-change response of men with good promotion prospects was that the computer's effects would be favourable. In the branches we had found that men who liked the 'interest' in their work were less approving of the computer than others; we did not find this in North Office but we did find that this group were critical in that they believed the computer would have a damaging effect on the clerical career prospects of girls.

In North Office then, although the reality of automation had not been experienced, the anti-automation clerk was essentially the same kind of person as in the E.D.P. branches; a man in his twenties, and probably late twenties, with good promotion prospects, identified with the bank and its objectives and very conscious of the importance of customer service. The pro-automation clerk also had the same characteristics in both situations: a young man, not greatly identified with the bank or banking, who told us that there was nothing much he liked about his job.

Three variables were common to North Office clerks irrespective of sex, and these again supported our hypothesis that identification with the bank and a belief that the status quo could lead to the attainment of personal goals led to opposition to change. All clerks who thought their promotion prospects were good were likely to see the new system as having undesirable consequences for the bank, while clerks whose promotion prospects were poor were more prone to see the computer's effects as favourable. Similarly, clerks who were over 21 were very much more likely than younger clerks to comment on the computer's disadvantages for the customer. Again, these attitudes are similar to those we had found in the E.D.P. branches. There was one difference, however. In North Office there did not appear to be any relationship between membership of a trade union and attitude to change.

This comparison of factors influencing attitudes to change in the E.D.P. branches and North Office suggests that it is not direct experience of the impact of change that influences attitudes so much as how an individual sees the consequences of an actual or possible change as affecting his goals and objectives. Therefore clerks with similar interests in the E.D.P. branches and North Office tended to react in the same way towards the computer. The principal difference was that whereas we had some evidence that the morale of male clerks had dropped in the E.D.P. branches, presumably because of their experience of the new system, morale in North Office had remained at its pre-change level. We must qualify our first statement, however, and say that it is probably only true if the change is relatively mild in its effects. If a technical innovation alters work in some particularly unpleasant way, as happened with the Royal Exchange machinists, then this is likely to produce highly critical attitudes in the affected group, and these attitudes may not be repeated in the non-affected group.

TABLE 18 *Variables affecting attitudes to change*

North Office anti-change variables	E.D.P. branches anti-change variables	North Office pro-change variables	E.D.P. branches pro-change variables
Liking the 'interest' of work	Liking the 'interest' of work	Liking 'nothing in particular' about work	Liking 'nothing in particular' about work
Good promotion chances	Hoping to get promotion	Poor promotion chances	No promotion plans
21-30 years of age (particularly 25-30 years)	21-30 years of age	Under 21 years of age	Under 21 years of age
	Member of N.U.B.E. or Staff Assoc.		Not a T.U. member

The handling of the change process

In most situations the way change is handled is of equivalent, or even greater, importance in conditioning attitudes, than the actual consequences or perceived implications of the change. If a new system is introduced easily and smoothly, staff are more likely to welcome it than if the transition period is accompanied by stress and turmoil. If, when problems do occur, these are overcome rapidly, and if change is presented in its most favourable light, with emphasis on its advantages, then staff will be more inclined to accept than to oppose it. In chapter 4 we showed how staff who heard about the bank's plans unofficially were less favourably disposed towards the computer than those who heard officially from management. Let us now see what the clerks thought of the change process.

PROBLEMS MET BY E.D.P. BRANCH CLERKS

Clerks who had experienced both the old system and the new were asked to describe the kinds of problem that had to be overcome during the change-over period, and the problems they still had to contend with. Most clerks had a great deal to say on these subjects and the following table shows the kinds of problem they referred to most frequently.

TABLE 19 *Problems that had to be overcome during the change-over period**

	King George's Place	St. John's Avenue	Paul Street	Ringer Square	TOTAL
Retraining problems	—	—	5	—	5
Organisation problems	3	1	5	4	13
Customer relationships	5	2	3	2	12
Technical problems	3	—	2	5	10
Working conditions	3	—	—	1	4
Communication	—	—	—	3	3
Miscellaneous	1	1	—	1	3
Don't know, no answer	2	—	1	—	3
	17	4	16	16	53

* Fourteen clerks gave more than one answer to this question.

Problems encountered during the change-over were fairly similar in all four branches, although Paul Street was the only branch referring to training problems, and Ringer Square the only one referring to communication difficulties in answer to this question. Paul Street had not received its punch machines when they were expected and this delayed and hindered training. Machinists commented 'the difficulty was trying to teach other people and practising one's own job at the same time', 'learning to work the machines was the worst. The girls weren't used to them and they kept breaking down because of this'. Ringer Square staff had some strong comments on communication, 'We didn't know what we were meant to be doing and this didn't become clear for several weeks'. A computer clerk felt managers didn't have enough information. 'People in authority get the wrong end of the stick because they have no knowledge of computer work. They don't know what can and can't be done on the computer.'

Organisational difficulties were referred to most frequently, particularly the necessity for re-arranging the work system to meet computer requirements, and the fact that—in Ringer Square in particular—this had to be done in a hurry and caused an excessive amount of work. 'There were thousands of problems—one was adopting the flow of work to the new system,' 'We had to modify our existing routine and at the same time amass data for the computer.' A Ringer Square clerk said with some bitterness, 'We were advised we were going on the computer over two years ago, then we didn't go on. In the end we were given five weeks notice and had to start from scratch again—we're still trying to catch up.'

Customer relations were seen as the next important change problem, and clerks again described their difficulties in selling the new system to customers. 'The difficulty was getting people to understand that they couldn't have a statement if they came into the bank and asked for one.' 'Getting customers to accept the change-over. Most have been against it. None have said they liked it.'

Technical problems came third; the need to personalise all cheques and number all accounts so that the computer could identify these, and the inevitable breakdowns associated with the new branch punch machines.

These three problems, organisational, technical, and customer relations, were mentioned far more than any others; women referring more to technical and customer relations problems, and men to organisation.

The different occupational groups had their own particular problems in changing over to the new system. Four of the five women supervisors spoke of difficulties in persuading customers to order statements in advance and to accept the non-narrative format. The two supervisors of machinists described how two or three of the new punch machines had broken down each day. 'The mechanics took a day or so to fix these, and often didn't have enough training and experience to repair them properly.'

The machinists had problems with the machines. 'The machines kept breaking down, tape was not punched correctly and kept coming back to be done again. In addition the computer was always breaking down which meant customers' statements were late in coming from the computer centre.' One machinist summed up the situation: 'We just didn't know what we were supposed to be doing and it didn't become clear for several weeks.'

This feeling of uncertainty and confusion was mentioned by other women staff. A cashier told us, 'No one knew what they were doing. We had to work things out as we went along to fit our own requirements—no two branches are the same'.

Male clerks had been more concerned with planning the re-organisation than female, and their comments frequently referred to this. They spoke of the need to standardise procedures, to modify the old routine, and to amass basic input data for the computer. Many referred to the rush and the heavy workload during the change-over, and to the difficulties staff had in understanding the new system. A number of the men seemed in the dark about what had taken place. One said, 'the computer chaps here knew what the problems were—all I knew was that we were working later and later every night.'

Three of the staff responsible for computer introduction in the Paul Street branch were interviewed and they spoke particularly of training and communication problems. One said 'I had to familiarise everyone with the new system, yet I didn't know enough myself to gen people up.' Another, 'there hasn't been any real training, people in authority get things all wrong because they've had no training in computer work. We had a lot of trouble with the last balance because we checked and re-checked things when there was no need to check at all.'

TABLE 20 *Problems still in existence**

	King George's Place	St. John's Avenue	Paul Street	Ringer Square	TOTAL
Technical problems	4	1	7	4	16
Customer relationships	6	—	3	5	14
Organisational problems	3	—	1	2	6
Working conditions	3	—	—	—	3
Communications	—	1	—	2	3
Retraining problems	—	—	—	1	1
Staff don't like new system	—	—	1	—	1
No problems	—	2	1	—	3
Don't know, no answer	—	—	1	—	1
	16	4	14	14	48

* Nine clerks gave more than one answer to this question. New starters were not asked it.

Once the change-over was completed, organisational problems diminished, but technical problems continued—particularly at Paul Street where nearly half the staff interviewed said this was their major current problem. At the time of the post-change interviews work at this branch was still being delayed by breakdowns at the computer centre. Customer relations were also a continual source of worry to staff, particularly at King George's Place and Ringer Square, although nothing could be done about this apart from going on persuading customers to accept the new system. The great identification of branch staff with the needs of the customer which came out so clearly in our survey is interesting, and indicates a sense of personal service which must be unusual in commercial organisations.

Lastly, the clerks were asked if they thought the change-over to a computer could have been carried out differently, and many had strong views on this point. Half thought it could have been, and the points most frequently mentioned were the need for more consideration by computer centre personnel of both the problems of branches and staff working in them. King George's Place staff felt they had suffered through being the first branch to go on the computer. 'We were guinea pigs and bore the brunt of the difficulties. It would have been better if another branch had been chosen.' One clerk there suggested, 'I think they brought the computer into branch banking a little too soon in view of the machine's limitations. Everything seems to have been carried out by a system of trial and error.'

Communication was now frequently mentioned, although more often by men than women. One woman supervisor thought it would have helped if someone from a branch already on the computer 'had come and worked with us for a few weeks and if, in turn, we had been able to spend some weeks in her branch seeing how the new system operated'. A machinist suggested: 'We should have had someone who understood the computer to tell us about it. We just had a pile of notes to read.' It seemed that even when information was given to staff the way in which this was done was not effective. The women clerks wanted verbal and visual information, not documents.

Men were more critical of communication deficiencies than women, and spoke of the need for more communication—both between the branches and the computer centre, and between one branch and another. Two of the branch computer officers thought there should have been more information from the bank's Head Office; 'people didn't know what the new system meant, and they still don't'. Other clerks repeated this criticism; 'we were working completely in the dark', 'we were thrown on the computer—not given sufficient tuition—left to sort too much out for ourselves'. 'The computer manager came over once in two years to talk to staff, and was never in the branch more than two or three times after that.' The need for greater liaison between the branches and the computer centre was also stressed. More information was seen as essential, but clerks thought this should have been combined with consultation and co-operation. Visits to the computer centre would have been welcomed and we learnt later that the O. & M. Manager did invite branch staff to go there. In fact, although it was so near the affected branches, few of the clerks had visited it, and many seemed to picture it as a kind of mad scientist's laboratory.

There seemed to be very little official communication between one branch and another during the change-over, and clerks criticised this. 'Other branches could have benefited from our mistakes if they had been told about them.' 'We needed some insight into what the computer was going to do, and branches already on could have told us this.' 'We were not given any practical experience, only a pile of notes to read. The whole office could have gone in turn to a computerised branch if we'd started the change-over six months earlier.' This last comment was made by a Ringer Square voucher clerk and seemed justified as this branch waited many months for its eventual transfer.

It is interesting that although such a long period of time elapsed between the transfer of the first and last branches, the handling of the change process did not improve with experience. The computer and O. & M. research staff responsible for this apparently did not know of the branch staff's criticisms, and made the same mistakes in the last branch as they had in the first. It can be argued that the change strategy should not have been left entirely in their hands as, not unnaturally, they directed most of their attention to technical and systems problems. Organisational and technical problems will always occur to some degree when introducing any major technical change, but communication problems can be avoided if steps are taken to ascertain in advance what staff need and want to know. The bank did not pay enough attention to this area of change and its staff were critical of it for not doing so. Head Office could usefully have involved itself more in these communication problems, and provided some of the information services that staff felt to be lacking.

COMMUNICATION AND THE NORTH OFFICE CLERKS

The North Office clerks were also asked to comment on the difficulties of the change period, the problems E.D.P. branches still had to cope with, and whether the change-over could have been handled differently. But it soon became clear that they had virtually no information on any of these points. Although some information was reaching them from the E.D.P. branches, it was not about these areas of the change process. However, as one of the objects of studying North Office was to observe the diffusion of information about the computer into non-affected branches, we had introduced three additional questions into our North Office post-change questionnaire. These were:

> 'How have you learnt about the computer and the new work system?'
> 'Do you feel you have been given enough information about the computer and how it is affecting the work of the bank?'
> 'If not, what other information would you like to have?'

The table below sets out the answers to the first question.

TABLE 21 *Source of information about the computer*

	NONE	Talk and gossip	Personal experience	Journals or lectures	TOTAL
MEN	5	9	2	2	18
WOMEN	10	15	3	1	29
TOTAL	15	24	5	3	47

It can be seen that little formal communication of any kind was reaching North Office. A third of the clerks said they had no information about the computer, other than the fact that their own accounts were now processed on it. A half said that what information they possessed had been picked up through talk around the office. 'I've only heard from other people I work with,' 'Only in general conversation,' 'Just gossip,' were typical comments. Five of the clerks had either visited or worked for short periods in the computer centre or one of the E.D.P. branches and so had some direct experience of the new system. One clerk had been present at a lecture given by the computer manager, two had read articles in bank magazines.

Only 8 of the 47 clerks felt they had received enough information from management about the new work system and several of these 8 replied with a qualified 'Yes'. 'Yes, I have, because I'm not interested,' said one control clerk. 'I personally think I have enough information,' said another who had visited the automated branches, 'but other people have not been told much, and they don't want to be—they're frightened.' Some of the clerks who said they had not received enough information had some strong comments to make on this: 'We've been given *no* information. Mr. Brown's (the computer manager's) lecture was five years ago, there's been nothing since then.' 'We've not been given any official information. We still don't know whether the new system will come to North Office.'

The table below shows the kind of information clerks said they would like to have received:

TABLE 22 *Information about the computer requested by the clerks*

	How it will affect us	How it affects the customer	How it affects banking	How it works	General information, more about it	NONE	No answer, don't know	TOTAL
MEN	2	2	3	2	1	2	6	18
WOMEN	6	—	—	4	11	—	8	29
TOTAL	8	2	3	6	12	2	14	47

Men were more specific in their information requirements than women, and more concerned with the computer's effect on the customer and on banking. Below are listed typical masculine comments, one from each category in the table:

> 'I'd like to know the future of North Office. No-one tells us when we're going on the computer or if we are. Banks don't commit themselves.'
>
> 'I'd like to know the reactions of customers at branches which are operating the new system.'
>
> 'I'd like to know what changes there will be in the bank. Whether we'll become indirect machine operators or stay as clerks. In twenty years there may just be mechanics and machines.'
>
> 'I'd like to know how the computer works and what it can do.'

Women, in contrast, spoke much more of their desire to know the implications of the computer for them personally, and they were also less precise in their information demands. 'We'd like to know more about the computer, how it will affect our jobs.' 'We'd like to have been told about it, because we know nothing.' 'I would like to know a lot more about it.' 'I don't know anything about it, therefore I want to know a lot'. Some of the women referred critically to the bank's approach to communication. A girl who had visited one of the automated branches said 'Very few people know what it's all about. Even I don't know and I've been at Paul Street. We didn't even know about the computer centre and it's just around the corner. It's seek and find in the bank.' A machinist suggested: 'We should have been told exactly what is going to happen by management. We've been told nothing'.

These comments demonstrate very clearly that good communication is as necessary in a branch or department that is not going to be affected by a systems change as it is in those sectors that will be affected.

Five years after the computer first arrived in the bank, North Office clerks were still trying to find out what the new system meant, both for themselves and for the bank and its customers. Despite the fact that a small section of the work of their own office was now processed electronically, they knew virtually nothing about how the computer worked. More important, they knew nothing about the bank's plans for their own office—would they eventually go on the computer or would the traditional system of banking continue for some time to come. Rumour and casual conversation were almost their sole sources of information. Through this they had arrived at a particularly unfavourable perception of the computer's effect on the bank and of the job changes that had taken place in the four automated branches. Yet both men and women believed they should have been given information and were critical of the bank for not doing so.

If Royal Exchange transfers the work of North Office to the computer at some future date, it seems probable that staff resistance there will be greater than in the branches because unfavourable attitudes have been allowed to develop unchecked.

CHAPTER 7

Carters—The introduction of a new system

Our second firm, Carters, was a medium sized firm in the cattle food industry. The original company had developed as a sugar refinery in the late nineteenth century and until the early 1900s it was concerned mainly with merchanting sugar and molasses. In 1909 it started to manufacture animal foodstuffs and this eventually replaced the refining side of the business. The firm expanded rapidly in the years preceding the second world war and in 1947 a second firm was absorbed into the first, thus forming the present company. Since then the firm has continued to expand and now has a national coverage of markets. Competition in the industry is considerable and there are a number of competitors in the firm's immediate locality. Despite this the firm is prosperous, it has a good reputation as an employer and is not troubled by disputes.

In order to understand how Carters came to get a computer it is necessary to go back to 1958 when the Chairman of the Board saw an article in an American magazine dealing with the application of linear programming to the mixing processes used by the firm. This article was passed to the Accountant, also a member of the Board, so that he might consider the possible application of these techniques in Carters and, before very long, preliminary approaches were made to a number of manufacturers in order to discover a computer which could do the job in mind at reasonable cost. While negotiations were still in their early stages it became clear to these Directors that the computer could also be used in the accounts department for the processing of customers' accounts.

The reasons for the decision to acquire a computer are, in this instance, therefore, somewhat complex and hard to disentangle. Although interest was aroused initially in the idea of using a computer for linear programming, this was quickly enlarged to cover the accounting procedures. The principal advantage of a computer in the accounts department was thought to be its effect on staffing problems. Part of the work in this section of the offices was carried out successfully by a Hollerith installation where staff problems were minimal, but much of the general office work was tedious and unpopular. Although it was not difficult to recruit labour, it was difficult to keep it and labour turnover was extremely high. There was a great deal of overcrowding in the offices, yet no possibility of further office expansion. The computer's potential for saving staff would therefore help both the labour turnover problem and ameliorate poor working conditions. An additional factor encouraging office automation was the character of the firm itself. The Chairman was interested in new methods and the firm had been a pioneer in the use of punched card equipment. The success of the existing punched card installation had prepared the way for further steps in mechanisation.

Once the firm had made its decision to invest in a computer it next had to find a make and model suitable for its purposes. Carter was looking for a relatively small machine and so it asked only three manufacturers for detailed feasibility studies. Investigations took about a year, including six months discussion and planning with the manufacturer of the chosen machine, and the contract was eventually signed in May 1959. The firm chose to rent its new machine rather than buy it outright.

Selecting and training personnel

The machine ordered, Carters now had the task of preparing for E.D.P. and, although the delivery date was some time away, there was a great deal to do before the computer arrived. The first step clearly was that of finding suitable staff to program and operate the computer. The job of selecting and training this staff was given to the firm's Accountant and this executive was faced with the task of building up a team of assistants who would be competent to analyse existing procedures and transpose these into a form suitable for the computer. This involved the preparation of flow charts, block diagrams and finally the machine code. A great deal of work had also to be done in preparing for the conversion from one system to the other, for example re-designing documents to meet computer needs and planning the sequence in which work procedures could best be transferred.

The computer manufacturer recommended the firm to select its programming staff from within the organisation, arguing that it would take much longer for programmers to learn the firm's system of work than for existing staff to learn how to programme. Carters, however, for a number of reasons, preferred to adopt a different course of action. The firm, because of its general recruitment policy, was staffed mainly by young girls with low educational qualifications, and management doubted whether sufficient talent existed for the internal recruitment of programmers. At the same time, the fact that the primary interest of the firm was in using the computer for linear programming meant that mathematical ability seemed particularly important. Accordingly, efforts were at first made to recruit a mathematician. In May 1959, when the computer was ordered, the Ministry of Labour and every university in the country were approached to see if they had suitable candidates, while advertisements were also placed in the national press. The result, however, was extremely disappointing. A few girls applied before they sat for their final degree examinations, but did not re-apply afterwards. One girl with wide experience in statistics was interested, but the salary she wanted, while justified by her experience, was more than the firm felt the job to be worth. Finally, a girl with a Bachelor of Commerce degree applied and was accepted, only to be turned down as unsuitable by the computer manufacturer after she had taken his programming course.

Meanwhile, the firm was beginning to realise that a mathematician was not necessary, especially as the computer manufacturer was now handling the linear programming side. Accordingly they advertised again, this time asking for clerical experience, rather than qualifications in mathematics. Two young men were now appointed, one in his middle twenties, the other almost thirty. Both had good G.C.E. 'O' level and School Certificate results and considerable clerical experience. One, the older of the two, and appointed as senior programmer had had several years in the Merchant Navy. Both were still studying part-time, one for the Institute of Cost and Works Accountants, the other for the Institute of Secretaries, and both had successfully completed the intermediate stage of these examinations. Neither had had any experience of computers. The senior programmer was appointed towards the end of 1959, the other in January 1960, and this gave them just over a year before the computer was expected to arrive. However, during this time the firm did have the loan of one of the computer manufacturer's programmers and he occasionally worked with the new programmers, who were also given five weeks training on the manufacturer's own programming course. Apart from this, they had to pick the job up for themselves. Responsibility for the computer installation was given to the accountant, but day to day supervision was handed over to the manager of the Hollerith installation.

This firm, with its small computer, did not require the large supporting staff associated with the bank machine and so did not have to recruit additional staff as punch girls and computer operators. Training was also less of a problem as a number of girls were accustomed to operating the Hollerith equipment and the new computer installation did not require very different skills. The Hollerith manager believed that special training in punching was unnecessary as the girls would adapt to computer requirements quite easily. He hoped that girls who were already experienced Hollerith operators would take over the task of operating the computer.

The installation period

The first major job for the programmers was to make a thorough analysis of work procedures in the Accounts Department in readiness for their transfer to the computer. Because invoice preparation and ledger posting were already completed mechanically they presented little difficulty. The main problems were the re-organisation of some ancillary aspects of the accounts work—notably pricing, credit control and the receipt and identification of payments from customers. Also the old system had involved three different sections, each with its own head, and for two of these sections some loss of function might be involved. The section head in charge of the Hollerith installation, now in charge of the new computer installation, would, on the other hand, have his responsibilities greatly increased.

The task of systems analysis and programming was primarily in the hands of the senior of the two programmers. An apparently able young man, he was nevertheless handicapped in his work both by his lack of experience of E.D.P., and by the fact that he was an outside appointment and so had to learn the firm's system from scratch. The computer manufacturer provided some assistance with programming but not the amount that Carters thought was needed.

The computer arrived in March 1961, and it was then handed over to the engineers for several months of testing while the two programmers continued designing programs. By July test programs were being run successfully, but the firm decided that it was premature to transfer any of the Accounts work, as July was the firm's holiday period and staff needed for punching were not available. Therefore the target date for the production of the first invoices on the computer became September 1st. It was hoped that all work scheduled for the computer would be successfully transferred within six months of this date.

Parallel running was started in September as planned, but the first of what was to be a long series of technical difficulties now occurred and the transfer of work had to be abandoned. The unfortunate programmers had to wrestle with the problems of machine breakdown while they were still unfamiliar with the procedures of normal computer operation.

Parallel running was eventually resumed in November, and in December the first phase of the move to an E.D.P. system began. By February 1962 the computer was dealing with one-sixth of the firm's retail invoices. This was not satisfactory progress but, during the period December to February, more difficulties had arisen due to computer breakdown and operating mistakes. In consequence work schedules had again fallen seriously behind. In order to produce the invoices to which they were now committed, the programmers had to work extremely long hours including evenings and weekends. However, even with these long hours, they could not produce invoices at the intervals required by the sales departments. A backlog of work accumulated which added to their day to day problems of getting the computer to run effectively.

Difficulties were also experienced with input reliability and here the firm's complicated pricing system proved a major problem. Prices were checked manually at first as the computer was not entirely trusted, but in every case of error it was found that the fault originated in incorrect coding, punching or programming, and not with the computer. This kind of error caused the computer to stop and the programmers had then to spend considerable periods of time identifying and correcting the wrong codes before work could proceed. Paper and carbon inadequacies also caused problems. The paper was inclined to break when invoices were printed out and the carbon in use during this period did not provide satisfactory reproductions.

The situation improved towards the end of February and great efforts were made to complete outstanding work. One-third of the firm's retail invoices were at last being successfully produced by the computer and phase two of the take-over plan was able to commence. By the end of May 1962 the entire retail side of the business had been transferred. Because work was being completed on schedule, more time was available for maintenance and a 15 minute period was allocated for this purpose each morning, together with one afternoon a week. Machinery ancillary to the computer was also working more satisfactorily although there were still difficulties with paper and carbons.

It had been hoped that wholesale invoices would be on the computer in May or June but, in fact, this transfer did not take place until the end of the year. Machine breakdown caused more trouble, and work began once again to fall behind. The computer manufacturer who had always tried to make out that technical problems were due to Carters' own negligence, now suggested that dust in the atmosphere might be a cause of breakdown. However, the computer staff were doubtful about the correctness of this idea, and they pointed out that other firms were able to run computers successfully in dirty environments. Their own suggestion was that, because the machinery was so frequently out of action, when it was working it was of necessity used for the maximum possible time. This led to a neglect of regular maintenance procedures and to a vicious circle; breakdown maximum use, breakdown.

By January 1963, nearly two years after the arrival of the computer, only two-thirds of the invoices scheduled for transfer were on the machine and there appeared little likelihood of the last third being transferred in the immediate future. The computer still broke down continually: the work of processing those invoices already on the computer was considerably behind and the firm was beginning to think seriously of scrapping the present machine and replacing it with a more up-to-date model. The manufacturer still maintained that dust was a major factor in breakdown and he suggested that the company install an air conditioning and refrigeration plant. Carters was, however, naturally reluctant to go to additional expense without any guarantee that plant of this kind would produce an improvement. The firm was now greatly dissatisfied with the technical reliability of the computer and there was a feeling that it had been sold a 'rogue' model.

In addition to technical problems, the computer staff were finding that the machine's drum was not adequate for all the price variations involved in the firm's marketing policy. An attempt was made to condense existing programs, but the computer manager suggested that if the number and variety of the firm's commodities continued to grow, for this reason alone the firm might be forced to replace the present computer by one large enough to handle all the data variations.

Conflict and strain in the computer centre

As we have seen, it was originally planned that the computer should take over the first data processing by September 1961 and a great deal of conflct developed between the computer centre and top management when technical problems and breakdowns began to disrupt the anticipated schedules. As one of the programmers put it at the time 'the Directors are breathing down the Accountant's neck, wanting to know when the computer will be ready. They do not like to see all that rent being paid out for something that is lying idle'. Because of this pressure from above, the Accountant in turn commenced putting pressure on the programmers to work harder and achieve faster results. The morale of the programmers, who were already working exceedingly hard, now deteriorated and they felt they had a number of justifiable grievances. One of these was the refusal of the Accountant to engage more staff to assist them in the routine work associated with getting the basic input program into the computer. The programmers believed that they were refused this very necessary assistance because the Accountant had sold the idea of the computer to the directors on the basis of reducing staff. He was, therefore, unwilling to negate his preliminary arguments by taking on any additional staff, even of a temporary nature. A second grievance was associated with financial rewards. The programmers had been promised a bonus for Easter 1962, but they believed that the very considerable extra work they had been compelled to undertake as a result of the breakdowns and consequent backlog of accounts for processing, warranted their receiving a small bonus at Christmas as a token of appreciation for their hard work and overtime. Instead of rewards, however, they maintained that they received only abuse and exhortations to work even harder. They believed that the accountant really knew in his heart the problems they were facing, but that he could not or would not admit, even to himself, that these problems existed, as this might have been interpreted as an indication of uncertainty concerning the correctness of the decision to get a computer.

By February 1962 the programmers were still both programming and operating the computer themselves, for although the intention was to train girl operators to take over this job, these future girl operators were still working on the Hollerith equipment. A number of factors were now making the programmers feel that their job was excessively onerous. Besides low morale, one of them at least was suffering from nervous and physical exhaustion resulting from the strain of continuous late night and weekend work. This led to a deterioration in concentration, and programming and operating mistakes were made which could have been avoided with less rigorous work conditions. As we have seen, technical breakdowns had been far more numerous than had ever been anticipated, and the programmers had to help the engineer cope with these, while testing programs and operating the computer. This heavy work load led to such errors as plugging faults—the incorrect wiring of the machine's plugboard—because the programmers did not have available the time necessary to master the technique.

One of the programmers now began seriously to consider his position and prospects with the firm and to see his future as lying elsewhere. He considered that he had been badly treated and that he was now in a position to use his new skills to obtain a more advantageous post in either programming or system analysis.

Although the morale of the programmers was particularly low during this period, the new computer manager, a man who had been with the firm many years, remained calm and unflurried. He maintained that he was quite satisfied with the progress that had been made so far. He had expected some trouble, and despite this he was still looking forward to the new system which he felt would be more streamlined and easier to work than the old Hollerith procedures.

In May 1962 relationships between the programming staff and the Accountant deteriorated even further. The programmers refused to do any more late working and as a result they became behind hand with the work. This led to a big row but also, as a result of this row, to some reduction in working hours through the introduction of a shift system. One programmer now came in at 7 a.m., the second at 9 a.m., and they took it in turns to take occasional afternoons off. With this system the computer was kept in operation from 7 a.m. in the morning until 8 p.m. at night. The introduction of shifts was, however, seen as a purely temporary measure by management. The Accountant was very much opposed to shift working as this involved the firm in extra rental payments and meant that the computer would at times be working without an engineer in attendance. The fact that one of the programmers had inadvertently burnt out the computer motor lent weight to this second point.

One of the two programmers now decided definitely to leave the firm and he secured another post in a different part of the country. Despite the conflict that had arisen out of the stresses of the situation he maintained that his reasons for the move were to some extent at least financial. He said that although he had been made a lot of promises his salary had never exceeded £800 per annum. In addition to this major source of discontent his personal relations with the Accountant had become very strained after he had made a serious operating error by putting an old price list on the computer. Fortunately this had been picked up by the Sales Department in the course of a spot check, but it had caused an already unhappy relationship to deteriorate further.

The second programmer also thought of resigning at this time, but the Accountant, realising that the computer could not function without programmers, offered him an increase of £400 provided he would agree to stay for another eighteen months.

The firm now decided to manage with one programmer only. Programming for the accounts department was nearly completed; in future there would be less work for programmers to do and the operating of the computer could be taken over by a girl. The manufacturer was also prepared to lend the firm one of its own programmers as a temporary measure to tide it over any short term staffing difficulties.

Relations between the computer centre and affected departments

In Carters very little information about the change was given to the staff and the process was carried out in an atmosphere of secrecy. There was no official notification that a computer was coming until October 1960 and by this time the systems analysis and programming had been going on for nearly a year. Moreover, even then only a broad outline was given without any indication as to the computer's ultimate effects. In fact, we were told that, outside the Board of Directors and possibly the Company Secretary, no-one had been told fully what the machine was intended to do except, of course, the programmers.

This secrecy was a matter of deliberate policy, and was justified by the firm on the grounds that greater frankness might raise difficulties not only amongst routine staff, but also amongst section heads who, the Accountant suggested, 'will worry if they know that they might lose work and so have their empires reduced.' Even several months after the actual arrival of the computer the sales directors had not been given full information on the consequences of the change.

Under the circumstances the question of consultation did not arise. The programmers were told to be tactful, not to reveal what they were doing, and simply to get information on the details of the current system. This made for considerable difficulties, and the programming staff met with suspicion and even, in one instance, hostility, as they tried to carry out their work. Not unnaturally the section heads in particular wanted to know what they were doing and why they were asking questions about the work.

This is not to suggest that those working in the department concerned did not know that a computer was coming. It is clear that most, if not all, knew of this at an early stage. They seem to have assumed, however, that its function would be to take over just the work of the Hollerith section and it was not generally realised that it would take over part of the work of other sections, notably Credit Control and Cash, and that eventually it would affect the work of the Sales and Traffic Control Departments. Once the programmers began to study the work of these other sections, fear and suspicion were aroused in the minds of some staff.

The programmers themselves disapproved of this policy of secrecy as, in their opinion, it made it more difficult to get the co-operation they needed. They believed the anxieties produced were worse than would have been the case if a policy of frankness had been followed. In order to combat this resistance they had to use a great deal of social skill and tact in selling the new ideas to staff, and they found that gradually co-operation began to replace hostility. Two sections of the Accounts Department, Credit Control and Cash, had always accepted the computer, but two departments, Sales and Traffic Control, had proved more resistant. Relationships between the computer centre and the sales department were eventually eased by the retirement of an elderly, conservative manager and his replacement by a go-ahead, much younger man. Similarly, relationships gradually improved with Traffic Control as the staff there realised the new system was not as unworkable as they had anticipated.

Whether departments were co-operative or not, however, a major source of difficulty in the early days was the production of faulty input data. For example, coding was now done in the Sales Department by clerks previously engaged on pricing. These clerks worked under pressure through having to provide data for the old and new systems of work simultaneously and, in consequence, they made many errors in coding. These errors caused the computer to reject the data and stop, and work was held up until the programmers had identified the incorrect code. The clerks objected to the requirements of their new jobs and said the work was impossible. Checking procedures were introduced in an attempt to improve coding, but errors still continued to be made.

Eventually the programmers evolved a training procedure for coders in an attempt to reduce errors. A set of very simple coding rules was devised and one programmer spent some time sitting with each coder in turn. He explained the new rules, checked their work and tried to persuade them of the advantages of the new system. Any coding errors that still occurred were carefully analysed in order to establish and remedy their cause. The clerks engaged on coding were relatively senior, labour turnover amongst them was not a problem, and it was hoped that an improvement in accuracy would result from this personal and systematic approach. However, as late as October 1962, technical breakdowns and coding were still the major problems of computer operation. All the coding was now checked, but errors still occurred. One difficulty, already referred to, was the large number of different prices associated with the product, and the necessity for coders remembering these. The firm was reluctant to standardise prices as their product was selling in a highly competitive market. Senior management believed standardisation would prove a disadvantage to the firm and advantageous to competitors, as the firm would no longer be able to offer price concessions to valued customers. There was a strong feeling that the computer should fit into the needs of the organisation. The alternative was to adjust the organisation to conform with the potential of the computer, and this was not thought to be an appropriate way of running the business. As it was the firm to a large extent now had all its eggs in one basket. If the computer broke down, the production of statistical data, invoices, etc., stopped at source whereas, with the old system, if one piece of equipment broke down the staff could still continue working on other clerical operations.

Relations with the computer manufacturer

An earlier chapter described how relationships deteriorated between the manufacturer of the Royal Exchange computer and the bank's computer centre personnel. A similar situation, for largely the same reasons, developed in Carters. Here again the manufacturer had given senior management time estimates for various stages of the take-over operation which proved to be wildly unrealistic. The computer staff found that the manufacturer had not made nearly enough allowance for technical breakdowns, for programming and for errors arising out of an imperfect system. The firm also found that the machine performance of both the computer and the ancillary equipment, particularly the punch machines, did not correspond to the manufacturer's sales talk. It seemed that this manufacturer too had oversold his product and the result was a loss of confidence in his equipment and a breakdown in manufacturer-customer relationships.

Conflict was heightened by the large number of machine faults and the attitude of the manufacturer to these. There was always considerable argument over who was to blame for these faults. The computer staff said the immediate reaction of the manufacturer was to accuse the user of the computer of negligence, whereas they were convinced that in fairness the blame should at most be equally divided between the two parties.

The computer staff also became irritated by the apparently contemptuous attitude of the manufacturer's representatives. They claimed that the latter acted disdainfully if asked for advice, or to rectify mistakes made by the firm's programmers. The programmers, not unnaturally, came to dislike asking for help. For a while the manufacturer had one of his own programmers in attendance at the centre, but by the time this person left relations were so bad that she and the firm's programmers were no longer on speaking terms.

In May 1962 the firm was so disheartened by the continuance of breakdowns that senior management had a full scale row with the manufacturer. As a result there was a descent of engineers, the equipment was thoroughly cleaned and protective covers were placed over the more delicate components. The manufacturer also replaced his resident engineer with another man who, although a trainee, was exceptionally keen and willing. In an attempt to combat accusations by the manufacturer that breakdowns were the fault of the firm's own staff, careful records were now kept of all operating procedures so that the programmers could demonstrate conclusively that this was not the case. But, despite these attempts at improvement, by January 1963 relationships with the manufacturer were as bad as ever. The computer manager and his staff were particularly incensed at a suggestion of the manufacturer that Carters should now rent one of their larger machines as the existing model was too small and the firm too ambitious in the amount of work it was attempting to process on it. The manufacturer's original feasibility study had stated that the work planned for the computer could be completed in five or six hours a day. In fact, it had proved impossible to complete it in less than eleven or twelve hours. In July 1963 Carters decided to scrap the existing computer and replace it with a larger model, but one of a different make.

This account of the installation of Carters' first computer is very similar to our earlier description of what happened in Royal Exchange; a disaster made up of ill chosen machines, technical problems, and an inability to meet planned time schedules. However, it is necessary to stress again the point we made in Chapter 1, that this kind of experience would be most unusual today; modern machines do not break down and simple accountancy procedures can be transferred in a relatively short period of time. But there are still many lessons that a manager installing a computer today can learn from Carters' experience. The most important are, first, the need for a great deal of preliminary research into the kinds of problems that can still be expected when installing a computer; second, the advisability of having one's own trained experts when embarking on any major systems change. The reliance on one or two amateurs plus the advice of an interested computer manufacturer seems to be asking for trouble.

CHAPTER 8

The accounts department

In this chapter we can examine the attitudes of clerks who were in a situation quite different from that found in Royal Exchange. Carters' clerical work was already partly mechanised through a punched card installation, and the group affected by the change was almost entirely female. We shall see whether these two factors exerted any influence on attitudes to office automation.

Before the change

Carters employed 147 clerks and these were distributed between Secretarial, Accounts, Sales and Traffic Control departments. Our interviewing was carried out entirely in 'Accounts'—as this was the first department scheduled to have its work transferred onto the computer—and we completed our pre-change study in December, 1960, three months before the computer arrived. We interviewed all the clerks employed in the Accounts Department as it seemed probable that the work of the entire office would be altered to some extent by the new system.[1]

WORK ORGANISATION

The Accounts Department was divided into three sections: one, Hollerith, dealt with sales invoicing and ledger posting, using the punched card installation; the second, Cash, handled cash and cheques sent in by customers and agents in payment for commodities, and the third, Accounts Collection, chased up unpaid accounts.

[1] In order that comparisons could be made between Royal Exchange and Carters, the clerks in both situations were asked the same questions on most topics. Differences in the questionnaires will be indicated as the responses to particular questions are discussed.

The Hollerith section received sales orders from the firm's traffic and agency departments; girls punched information from these on to Hollerith cards which were then processed to obtain figures giving the total weight and price of the order and the customer's discount, and to produce an invoice. Ledger posting, the other main function of this section, consisted of keeping customers' records up to date by the daily punching of information concerning payment for goods received. This was transferred to ledger cards which were kept for each customer. The Hollerith section also produced monthly sales statistics and a monthly list of balances, invoices, and debit and credit notes. The Cash Department checked off cash received against the customer's ledger card, to ensure that the amount sent in coincided with the amount owing, and issued receipts. Accounts Collection checked the ledger cards once a month and sent out appropriate demand notes. These letters contained requests of increasing urgency according to the length of time that had elapsed since the customer received the goods. A monthly summary was made of accounts over four months old, and this was produced from information supplied by the Hollerith equipment.

GROUP CHARACTERISTICS

At the time of our pre-change study, Carters' Accounts Department had a clerical staff of 34 excluding the three section managers. In contrast to the bank it was very much a female group, for the only men were the section heads and three young boys. As in the bank, however, the clerks were very young—two-thirds were under 21 and only 6 over 25. Few of them had been with the firm very long and almost a half had joined the previous year. Only 4 had held more than two previous jobs.

As a group they were very different from the middle-class bank clerks. Predominantly working-class in social origin, only 6 had been to a grammar school and only one girl had 'O' level G.C.E., although nearly a half had taken some kind of full time or part time commercial course after leaving school. Their reasons for taking up clerical work were more negative than those of the bank staff—a third said they had become clerks because they did not want factory work or because they were not trained for anything else. 'Clerical work was the only thing I could do. I couldn't be a demonstrator or a saleswoman'. None of these clerks were members of clerical trade unions and so we had to abandon our attempt to see if office automation stimulated union membership.

ATTITUDES TO WORK

Thirteen of these 34 clerks were working on the existing Hollerith installation (8 as machine operators or punchers, 5 as pulling clerks), 4 were typists, 2 comptometer operators, and the rest were on general clerical duties. The firm expected any redundancy that occurred as a result of the computer to be amongst those on general clerical duties and in typing, as they represented the 'traditional' type of office work which it was hoped the computer would supersede. The girls on the Hollerith installation would simply switch over to similar duties in connection with the computer itself. No redundancy was expected in this group, and no dramatic change in the nature of their work.

The majority of these girls seemed to have little positive identification with the tasks they were responsible for. When asked what they liked most about their jobs, only a third referred to some aspect of the work, such as 'its interest'. A third said 'don't know' or 'nothing', and a fifth said they liked the social relationships arising out of the work situation—'the other girls are very nice', was a typical comment.

We tried to assess their job satisfaction but encountered the same difficulty we had had with the bank clerks—it is not an easy thing to measure. To increase our problem two of the indices of satisfaction used in the bank could not be used in Carters. It will be recalled that in the bank we took liking the 'interest' of work, good promotion prospects, willingness to embark on the same career again, and desire for sons to take it up, as indications of satisfaction with work. But, in Carters, only 3 of the clerks wished their sons to take up clerical work—the girls were almost unanimous in favouring skilled manual work—and only 5 thought their personal chances of promotion were good. However, the answers to the questions 'if you were starting your career again would you become a clerk?' and 'what do you like most about your job?' still seemed relevant and we were also able to use length of service as an additional index. This had not been possible in the bank because the movement of staff from one branch to another made inter-branch comparisons impossible.

In Carters, work group seemed the most appropriate method of categorisation. With only three men it would have been of little value to compare sex differences in attitude. Also a breakdown by work roughly co-incided with a breakdown by section, as all the Hollerith staff were located in the one section.

TABLE 23 *Indices of job satisfaction*

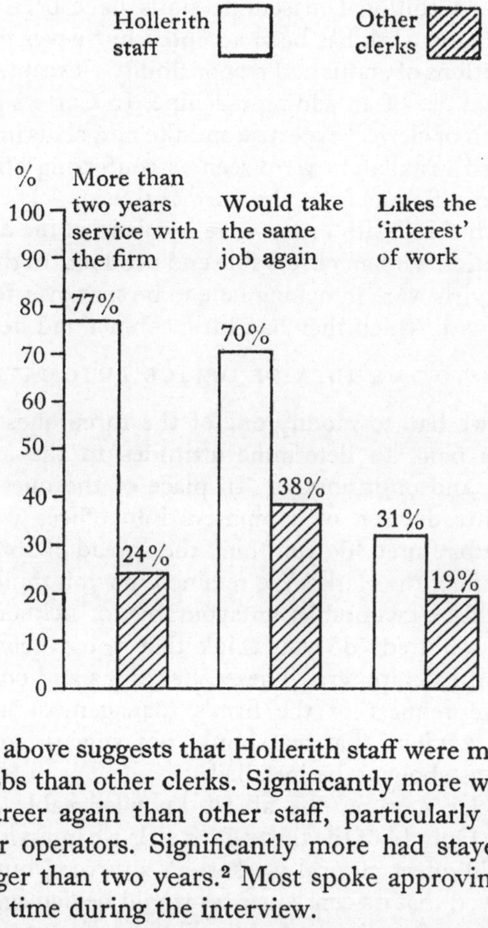

The table above suggests that Hollerith staff were more satisfied with their jobs than other clerks. Significantly more would choose the same career again than other staff, particularly typists and comptometer operators. Significantly more had stayed with the firm for longer than two years.[2] Most spoke approvingly of their jobs at some time during the interview.

This liking for machine work in Carters was combined with a belief that it was of higher status than most other clerical jobs in

[2] The significance is only at the 10% level.

the firm; the majority of those girls saying they would like to change their jobs expressed a preference for a move to the Hollerith installation. These attitudes are in strong contrast to those in the bank where machine work came near the bottom of the status hierarchy. It suggests that girls do not necessarily always dislike mechanical jobs such as punch operating and that the status a job acquires depends very much on other variables in the situation. In banking, the traditional masculine skills have been mental and clerical, machine work has been accepted but never performed by those in positions of status and responsibility—except, perhaps, for the occasional use of an adding machine. In Carters there was no such tradition of clerical expertise and the new skills involved in the punched card installation were seen as conferring status because they were so different from the normal routine office work. Also, the girls in the Hollerith group were involved in the completion of a total operation with an observable end product. In the bank, after the change, girls were punching tape to be sent over to a computer installation with which they had little contact and never saw.

ATTITUDES TO THE IDEA OF OFFICE AUTOMATION

In Carters we had to modify one of the three questions we had used in the bank to determine attitudes to the idea of office automation, and omit another. In place of the question 'do you think the introduction of computers into offices will affect the clerk?' we substituted 'do you think the spread of automation into offices will affect the clerk?' We retained 'do you think, in general, the effect will be favourable, unfavourable, or neither one nor the other?' but omitted 'do you think that you personally will be affected by the computer?' These alterations and omissions were made at the request of the firm's management as they were particularly anxious that we should not suggest to staff that a computer was about to be installed.

Carters clerks were not at all knowledgeable about office automation. Only 14 of the thirty-four girls we interviewed thought that it would affect clerical workers in any way, but half of this group believed that its consequences would be damaging, and they unanimously referred to the likelihood of redundancy. The majority of those expressing apprehension about redundancy worked on the Hollerith installation, an indication, perhaps, that this group were more aware than other clerks of the implications of automation and of a machine's potential for taking over human work. 'Yes, I think Hollerith and machines like these will eventually take over clerical work', said one girl. 'Machines are going to take the place of a lot of clerks' commented another.

Although the Hollerith group were aware of the possibility of redundancy, when asked the wider question, 'Will the effect of computers be favourable, unfavourable, etc.?' they were slightly more likely than other clerks to say favourable. This seemed to indicate either that they did not see themselves but the more traditional kinds of clerks as the victims of redundancy, or that they were speaking in terms of advantages for the firm, not for the individual clerk.

These replies show clearly how little the clerks knew about computers and office automation at the time of the pre-change interviews. Less than a half thought that automation would have any effect on clerks and, although the clerks knew by this time—largely through runour—that a computer was expected, there was no indication that they were anxious about its coming or that redundancy was widely expected. Moreover, those referring to redundancy were more likely to be from the Hollerith installation where no redundancy was anticipated, than on general duties, where a reduction in staff was planned. Up to this point, therefore, three months before the computer arrived, the firm's secrecy policy, although not effective in preventing news of the computer from reaching staff, had not apparently caused staff to worry—even though they were quite aware that something was being kept from them. It remains to be seen whether this atmosphere of calm prevailed during the change-over period when staff at last appreciated that they were involved in a major systems change.

FACTORS AFFECTING THE CLERKS' ATTITUDE TO THE IDEA OF CHANGE

It is not easy to identify these in Carters as we only have a small group of 14 admitting to the likelihood of change. Let us start by seeing if there are any variables distinguishing clerks who said 'yes', automation would affect the job of the clerk, from those who said 'no' it would not. The first one that stands out is experience with, and knowledge of, mechanisation, for Hollerith girls were significantly more likely to say 'yes' rather than 'no' to this question. Those girls saying that 'interest' was the thing they liked most about their job (not necessarily the Hollerith group) were also more likely to say that clerks would be affected by automation, presumably because, like the bank clerks, they were more identified with their work than other groups and more sensitive and resistant to the idea of change.

It was difficult to identify any characteristics apparently influencing a favourable or unfavourable attitude to change. However, clerks saying their jobs were interesting were also more likely than others to say automation would have unfavourable consequences for the clerk—they liked their present jobs and did not want them to be altered. Again, girls with formal qualifications, as we had found in the bank, were less likely to say automation would be favourable—they had more to lose from a new system of work that required mechanical skill, not commercial knowledge or intellectual ability. In contrast, girls saying they would choose clerical work if they were starting again were more likely to think a computer would prove advantageous. 'There will always be a need for clerks, and computers will make life easier'. Such conclusions as can be drawn from these answers suggest what one would expect—that clerks who are knowledgeable about mechanisation and do not feel threatened by it will, like the Hollerith girls, see an extension of it as a good thing. Clerks who have qualifications which are irrelevant to the new system or who like their present tasks and find them interesting will wish to preserve the status quo and see change as a threat. The fact that clerks who say they would choose the same career again are more likely to favour automation is more difficult to explain. This group were, however, significantly more likely than other clerks to want to change their jobs, and this suggests that while they were identified with clerical work generally, they were not so attached to their present jobs as to make them wish to resist change.

The three variables shown in American studies as correlating with a pro-change attitude, better education, middle-class background, and middle- or upper-class identification, again were not supported in our results. There was a tendency for girls who had attended secondary modern schools to favour change more than girls who had gone to grammar schools, but the grammar school girls were so few in number as to make comparison difficult. There was no difference between the attitudes of girls with fathers in manual occupations compared with fathers in white-collar occupations. Similarly there was no difference in attitude between girls who said they were working class and those who said they were middle class.

What little evidence we have from our two case studies appears so far to suggest the opposite to the American conclusions—that better educated girls are more likely to be against, rather than for, automation, because they see it as making their knowledge and qualifications redundant. However, a possible explanation for the discrepancy in the American and British results is that the American team, although studying office automation, asked questions about change in general. Our questions were specifically about automation and, therefore, directed at a particular kind of change with known or suspected disadvantages.

In Carters our pre-change data suggested that the existence of a mechanised accountancy system would make the impact of office automation less dramatic than in the bank. It also seemed likely that the Hollerith group would see a computer as an extension of their already well liked machine environment, and that they would not react against the new key punching jobs in the way the bank machinists had done. In fact, punching for a computer would be very similar to punching for Hollerith equipment.

The really interesting point emerging from the girls' attitudes, however, is the small amount of anxiety shown by the clerks despite the firm's unconventional communication policy. Personnel management text books always stress the need for adequate communication if a management wishes to secure the confidence and co-operation of its employees. Here was a firm that had adopted the opposite approach and decided 'we will tell them nothing' and, worse still, had not succeeded in keeping its secrets, for the news had leaked out unofficially. Personnel management theorists would predict confidently that staff in such a situation must be in a state of considerable alarm, yet Carters' clerks showed no sign of this. Why? There seem to be two likely explanations. First, the existence of a mechanised clerical system meant that girls in the Accounts Department were accustomed to machines, and we have already shown how work in the Hollerith team conferred status and was much sought after. News of a computer, however obtained, would imply an extension of mechanisation, and perhaps more opportunities to do this kind of work. It would therefore evoke approval rather than dismay.

Second, we have shown in our analysis of Royal Exchange attitudes how girl clerks on routine jobs are not involved in what happens at work; their attention is fixed on more important things such as leisure pursuits and matrimony. If their job situation deteriorates or they become too bored, they can easily move on to a new job and a new group of people. Why then, should they be alarmed at the idea of a computer?

These conclusions show that, as an attempt to avoid clerical anxiety, Carters' secrecy policy was unnecessary in the Accounts Department; although if management had kept these clerks informed, this would not have altered their attitudes. However, we cannot generalise from this that the absence of communication in Carters did no damage whatsoever. In the last chapter we described the programmers' difficulties in obtaining co-operation from departments, such as Sales and Traffic Control, who had to supply input data for the computer. Although we did no systematic interviewing in these departments, it was clear that clerks and supervisors there were hostile to the new methods of work. We can say that, in Carters, nothing was gained from the secrecy policy and a greater frankness might have increased acceptance in departments other than Accounts.

After the change

In any ongoing study of change, it is always difficult to select any one point in time and decide that this is an appropriate moment at which to study the consequences of innovation. Change is not something that happens and is completed, it is rather a continuing dynamic process. At Carters the post-change phase of the research was carried out in October 1963, nearly three years after the pre-change interviewing programme. This date was selected, not because it was thought the transition period was over, since the work of the Accounts Department had by no means been completely transferred to the computer, but because the firm had now definitely decided to scrap its first machine and replace it with a larger model of another make. This second computer was not due to arrive until May 1964, and its installation would necessitate a considerable re-organisation of the computer system. In the meantime the work of the Accounts Department was to be removed from the original machine and processed by another firm in the area which already possessed a machine of the kind Carters was proposing to install. Had the post-change study been left until the work of the Accounts Department was entirely transferred to E.D.P., this would have meant waiting for the new computer to arrive, for it to become operational, and for work not yet processed on the first computer to be transferred to the second. This would have drastically delayed the completion of the research.

WORK ORGANISATION

Before any staff were interviewed the Accounts Department was again observed in order to discover which operations had changed as a result of the first computer. We found that part of the work of the Hollerith section was now completed in the new computer centre.[3] For example, sales notes now went straight to the computer and data from these were transferred on to punched cards. These punched cards were sent upstairs to the Hollerith department and here they were collated into batches, sorted into account number order and merged with other cards giving the customer's name and address. Finally, balance cards showing the amount owed by the customer were selected from the records. All the cards were now ready for the computer. They were sent down to the computer centre once again, processed, and invoices printed out. The computer also produced ledger cards and accounts records. It had, therefore, taken over invoicing and ledger posting functions previously completed by the Hollerith equipment. However, because of the technical difficulty associated with getting the computer to run without breakdown, plans were a long way behind schedule, and, as we have seen, the computer was only able to process two-thirds of the invoices, leaving one-third to be completed on the Hollerith. Running the two systems side by side in this way imposed a great strain on the department as the girls had difficulty in coping with the amount of work involved.

The Cash Section of the department was only very slightly affected by the computer. The work of checking cash received with records of cash owing, and the issuing of receipts, remained as before, but a control function had been added to check that the computer did not produce invoices for goods that had already been paid for.

[3] In Carters the computer was sited in a downstairs office and not in a separate building.

'Accounts Collection' had had its work load lightened to some extent by the computer, as statements were now produced by the machine instead of by typists. The Section Head indicated the month he required statements for and the computer was then programmed to produce these automatically. Ordinary statements were sent out to customers for bills incurred within the most recent month. Sharp reminders were incorporated in statements relating to bills more than a month overdue.

The staff of the Accounts Department was now 33, a decrease of only one from the pre-change number. Supervision remained the same, with a separate head for Hollerith, Accounts Collection and Cash, although the head of the Hollerith section was now also Computer Manager and responsible for the computer centre under the direction of the firm's Accountant. In addition, the firm had a full time programmer, and a part time engineer supplied by the computer manufacturer. It can be seen that so far no saving in staff had resulted from the introduction of the computer.

Although total staff numbers in the department were almost unchanged, the content of part of the work was found to be different from the pre-change situation and new functions had been introduced. These were the jobs of coder—translating data from sales notes into code for punching purposes—computer operator, and punch and verifier operators for the computer ancillary equipment.

Staff working on the Hollerith installation had been reduced from thirteen to nine; there were now six girls operating, punching or verifying, two card pulling and one checking work as it came off the machine. The computer had a work group of eight, one operator, one puncher, one verifier, two coders, and three invoice checkers. Although data processing was divided between the punched card equipment and the computer, the total number of girls employed full time on machines, excluding typists or comptometer operators, had remained virtually the same (eight pre-change, nine post-change). The number of typists had been reduced from four to two, a result of the automatic production of invoices and statements by the computer. Surprisingly little displacement had occurred amongst staff engaged on routine clerical duties, although this was the group originally thought most likely to be affected by the change. The number was now twelve instead of fifteen.

As we have seen, the Department's work load was increased by the simultaneous operation of Hollerith and computer systems, but the Computer Manager did not envisage any major reduction in staff numbers within the Accounts Department when the new computer took over. He suggested that card pulling would become redundant, but other jobs were unlikely to be affected. It would appear that even if the first computer had been in full operation it would not have resulted in any considerable staff saving. This is not to suggest that staff saving does not result from computer usage. In most circumstances it would, but Carters was using its machine in a very limited way.

Group characteristics

All thirty-three of the staff in the Department were interviewed for the post-change phase of the study. Seventeen of these had been interviewed in the pre-change programme; of the remainder, thirteen had joined the firm since the computer was installed and three had been working outside the Accounts Department when the pre-change interviewing was carried out. This time seven of the clerks were men, an increase of four on the pre-change number.

The thirteen new clerks had similar characteristics to the original group; they came from working-class homes, all except two had been to secondary modern schools, half had had some kind of part-time or full-time further education, but only four had any formal qualifications. Only one girl had a positive reason for becoming a clerk—she had taken a course at a commercial college and got her job at Carters with its help. The remaining twelve had all drifted into the work; 'I didn't know what to do when I left school', 'I didn't want factory work', 'I was sent by the Labour Exchange'. These answers were in sharp contrast to those given by bank clerks who had usually had positive reasons for their choice of career.

The job satisfaction of the new clerks did not seem particularly high. Only two suggested that they would choose the same job if they were starting their careers again—significantly fewer than in the pre-change study—and only two said 'interest' when asked what aspect of their job they liked most. It was not possible to identify any general change in the job satisfaction of those clerks who had been with the firm before the computer arrived. Rather fewer of the Hollerith and computer staff mentioned 'work interest' than before, but the pre- and post-change differences were small. Surprisingly, clerks were now, if anything, more pessimistic about their promotion chances than they had been before the arrival of the computer. We had expected the opposite—that staff would see the new computer centre as providing increased job opportunities—but, following the bank pattern of response, this did not happen. One man and two girls thought their promotion chances within the firm were good; the majority said their chances were poor or non-existent. There are two possible reasons for this increase of pessimism. In 1960, knowledge of the advent of a computer could have led staff to believe that there might be an increase in promotion opportunities, although these opportunities never materialised, but a more probable explanation is that there was a growing realisation that the computer was more likely to take away than to provide job opportunities. However, the change is a very small one and quite likely to have come about by chance.

TABLE 24 *Assessment of promotion chances*

	GOOD	POOR	NONE	Not interested or don't know	TOTAL
MALE	1 (1)	4 (1)		2 (1)	7
FEMALE	2 (4)	12 (8)	11 (7)	1 (2)	26
TOTAL	3	16	11	3	33

Pre-change responses are in brackets.

The 20 clerks who had known both the old and new systems of work were asked how the firm had been affected by the computer. Nearly all had something to say about this, although it was clear that girls working on the computer or Hollerith installations had felt the impact of the new system most of all. The majority of comments were that work could be completed more quickly since the introduction of E.D.P. 'We are doing the same work as we were before, but it is more up to date now than it has ever been. Invoices are going out quicker.' 'The firm has become more efficient in sending out statements. This used to depend on girls typing them and if one girl was off it made a lot of difference.' Others referred to organisational changes; 'the department has split in two, we're upstairs and the computer girls are downstairs'. Comments were neutral and approving rather than critical, although one girl referred to the computer's tendency to make arithmetical mistakes, 'it adds things on it shouldn't'.

All clerks were asked the more specific questions on the computer's principal advantages and disadvantages for the firm. The machine's advantages were again overwhelmingly seen as a speed-up in the rate at which work could be completed. But many of the comments on this incorporated some reference to staff reduction or staff saving. 'It does the job a lot quicker, but knocks the amount of staff down.' 'The firm's work has increased and is being done more quickly yet no extra staff have been taken on. Therefore, the computer has saved on staff.'

There was not the same consensus on the computer's disadvantages. Over a third thought the machine had no disadvantages for the firm and a quarter were not sure whether it had any disadvantages or not. The remainder of the clerks referred to the possibility of redundancy—although this was certainly not a disadvantage from the firm's point of view—to a deterioration in working conditions, especially increased noise, and to problems arising out of machine breakdown, increased work centralisation, and lack of flexibility in the system. 'So much work goes through the computer now, if it breaks down the whole system is stopped. With the old system we could always get some of the work out'. One girl told us 'if the machine stops, I stop'. This lack of flexibility of the machine compared with the human individual seemed to be one of the computer's principal disadvantages, and it will be remembered that Carter's computer manager had earlier referred to the problem of having 'all our eggs in one basket'. The computer could work more quickly and more accurately than the clerks, yet if it stopped the work system came to an abrupt and complete halt; there was no possibility of continuing the work manually.

THE COMPUTER'S EFFECT ON WORK

Not all the girls had their jobs altered by the computer, and we found that some of the clerks on general duties were doing exactly the same work as three years earlier. However, almost a half of the girls said their work had been changed to some extent and nearly all of them preferred their new duties to their old, although for a variety of different reasons. Girls previously on Hollerith equipment, but now operating and punching for the computer, found their new job easier and more straightforward; while girls still on Hollerith said the introduction of the computer into the system had made their work more interesting and varied. Invoice checkers said it was easier to check work produced by the computer than to check the work of typists; and typists and comptometer operators found they had less work to do. Only two girls were not entirely approving of their new jobs; one was being trained as a tabulator operator and had only been learning the job for eight days, the other was responsible for preparing work for the computer. She said her work now had to be much more accurate as the new procedures made it less easy to correct errors.

We found that after the new system came into operation more girls wanted to change their jobs than before and there was considerable preference for work on either Hollerith or computer equipment. Thus machine jobs were even more highly favoured by women staff in the department than previously; very few girls who wished to change specified anything else. It was of little consequence whether these jobs were attached to the computer or the Hollerith installation, as the prestige lay in the responsibility for operating a machine. We also found that none of the girls on machines or punching wished to transfer to non-machine jobs, although some were anxious to move up the machine hierarchy; for example, a Hollerith Tabulator operator wished to become a computer operator and two punch girls wished to become Tabulator operators. One punch girl expressed the opinion that 'punching is the best job the firm has, everyone wants to do it'. It seemed that the youth of the clerks in this firm, the elementary nature of their clerical duties, the lack of promotion opportunities, and the length of time the Hollerith installation had been in use all helped to enhance the prestige of machine functions.

The clerks were asked how the computer had affected the 14 different aspects of their jobs. Their replies are set out below:

Only 11 of the 20 clerks who had known both pre- and post-change systems said their jobs had been affected to any considerable extent by the computer and these were all girls. The three men in this group were still on traditional clerical duties and were quite unaffected by the new system.

The most noticeable change was in the amount of work that individual girls were required to do. Nine said work had decreased

TABLE 25 *The effect of computers on different job aspects**

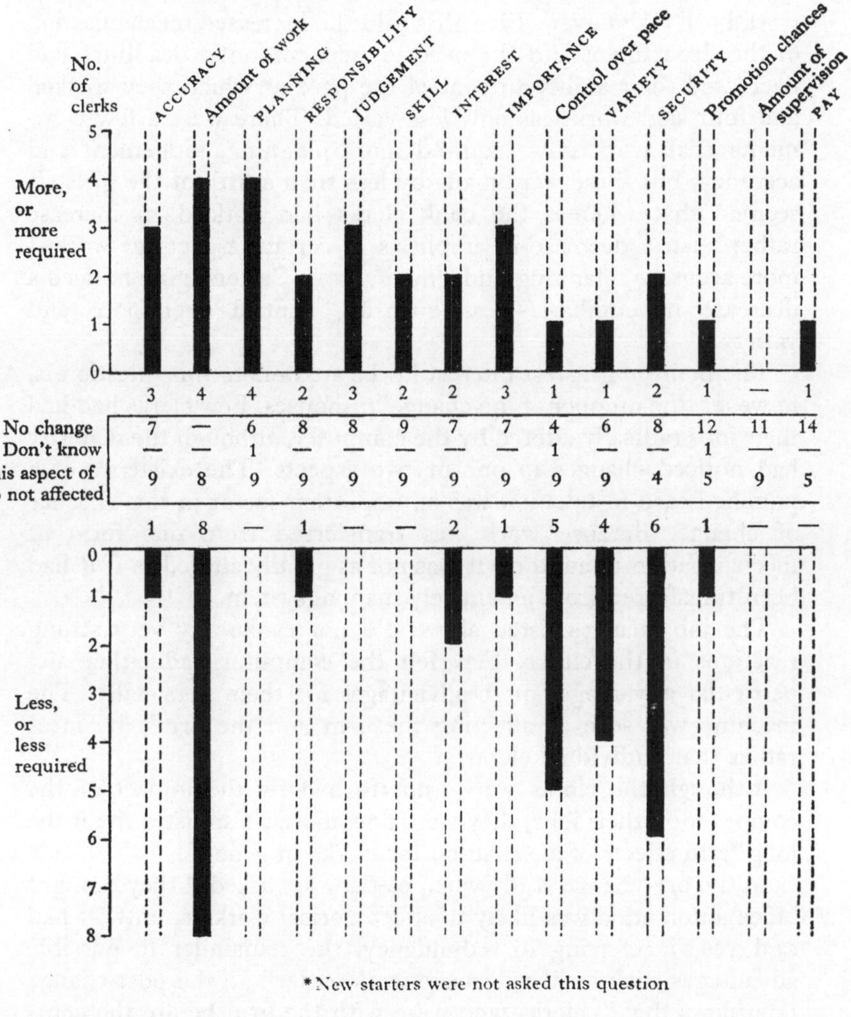

	Accuracy	Amount of work	Planning	Responsibility	Judgement	Skill	Interest	Importance	Control over pace	Variety	Security	Promotion chances	Amount of supervision	Pay
More, or more required	3	4	4	2	3	2	2	3	1	1	2	1	—	1
No change	7	—	6	8	8	9	7	7	4	6	8	12	11	14
Don't know	—	—	1	—	—	—	—	—	1	—	—	1	—	—
This aspect of job not affected	9	8	9	9	9	9	9	9	9	9	4	5	9	5
Less, or less required	1	8	—	1	—	—	2	1	5	4	6	1	—	—

*New starters were not asked this question

as a result of the computer, while four said it had increased. The next most remarked on change was in security of employment. Six girls thought this had been reduced by the computer and a typical comment was 'most of us will become redundant when the work is all taken over'. Five girls said the increased mechanisation of the department and the need to meet computer deadlines had decreased their ability to control the pace at which they worked and four said work was now less varied. There were a few comments that work now required more planning, judgement and accuracy, but these were made by less than a fifth of the girls. It seemed that whereas the bank clerks had noticed an increase rather than a decrease in emphasis in certain aspects of work—more accuracy, planning, judgement, etc.—Carters' girls noticed a decrease in emphasis—less quantity, control over pace and variety.

The point of greatest interest in the answers to this question is, however, the number of 'no change' responses. Few clerks had had their jobs radically altered by the computer, although the majority had noticed changes in one or two aspects. The existence of a punched card installation was an important factor in this absence of change. Because work was transferred from one form of mechanisation to another, it was not as greatly altered as if it had been transferred from an entirely manual system.

The job changes listed above did not evoke any very strong reactions in the clerks. Few felt the computer had either any particular advantages or disadvantages for them personally. The machine was seen as affecting the firm and the firm's business rather than individual clerks.

Although the clerks were unperturbed by the impact of the computer on their jobs, they were now more pessimistic about the long term effects of automation for clerks in general.

In the pre-change study when clerks were asked if they thought office automation was likely to affect clerical workers, only 14 had said yes, 7 referring to redundancy, the remainder to possible advantages such as 'it will make work easier'. In the post-change interviews the 20 clerks who were with the firm before the computer was installed were asked a general question on how they thought the introduction of computers into offices had affected clerical work as a career (a) for girls and (b) for boys. Their replies are shown below.

TABLE 26 *The effect of computer on clerical work as a career*

	For the worse	Made no difference	For the better	Don't know	TOTAL
(a) *For girls*	9	4	4	3	20
(b) *For boys*	1	5	4	10	20

As the clerks were mainly girls they were more prepared to comment on the effect of computers on girls and it can be seen that, like the bank clerks, their post-change attitudes were more pessimistic than their pre-change. Now nearly one-half of the clerks who had experienced both pre- and post-change conditions were pessimistic about the effects of computers on career prospects, whereas previously just over one-fifth of the Accounts Department had referred to likely disadvantages of automation. A number pointed out that computers did cut down on the amount of work available and that this had already happened to some extent in their own department. The girls seemed to be nervous about the future, although their own firm's computer had not at that time caused any staff redundancy at all.

Let us now examine in more detail the impact of the change on the different occupational groups.

In the post-change study we found that those girls who were in the new computer group had experienced the greatest job changes, but even in this small group there was no uniformity in the way work had altered. For example, punch and computer operators said they had less work to do but there had been no change in its variety—whereas girls checking invoices as they came off the computer had more work but less variety. There was no evidence that the group as a whole thought work had decreased in interest—the strongest impressions we had received from machinists in the bank. On the contrary, the computer group were highly approving of automation and its consequences. One told us 'most girls like to work with computers. All the girls who come here have this as their aim'.

The Hollerith team, still coping temporarily with work the computer had not been able to take over, had not been greatly affected by the changed work system, except that they now had less work to do. They, too, tended to approve of the computer, whereas the general clerks showed little enthusiasm. The computer and Hollerith girls were also convinced that the computer had no disadvantages for the firm, whereas few general clerks said this.

Surprisingly, however, all those girls suggesting that their security of employment had deteriorated were in the computer and Hollerith groups. A typical comment was: 'Most of us will become redundant when the work is all taken over'. No general clerks mentioned this although, as we remarked earlier, it was this group the firm intended to cut down. The explanation could be the same as in the pre-change interviews. The computer and Hollerith groups were more aware of the labour saving potential of automation and, in addition, they could see for themselves that their own only partially effective computer system was already speeding up and cutting down work.

Typists and comptometer operators were slightly affected by the computer as the machine had taken over some of their functions, but general duty clerks had experienced virtually no changes in their individual jobs. However, they did get the back wash of some of the technical difficulties and they were more inclined than other staff to comment on the machine's deficiencies. 'We have trouble when it makes mistakes or breaks down.' 'Sometimes it adds or takes away money; when this happens we have to correct the mistake by making a journal entry.' Despite their criticisms, over half this group thought the computer had advantages for the firm, particularly the speed at which work could now be completed.

FACTORS INFLUENCING ATTITUDES

In our analysis of attitudes at Royal Exchange we used the answers to four questions as an indication of pro- or anti- attitudes to change. These questions asked the clerk how the branch or bank had been affected by the computer, what the advantages and disadvantages of the machine were for the bank and for the individual, and how computers had affected clerical work as a career for girls and boys. By correlating the clerks' answers with their social characteristics and attitudes to work we were able to obtain a profile of the kind of man who, in banking at least, is likely to oppose rather than welcome change. Let us now see if we can do anything similar at Carters, bearing in mind that we were unsuccessful in identifying any variables influencing the post-change attitudes of the female bank clerks.

Our pre-change analysis indicated that, at Carters, certain factors went with a particular attitude to the idea of change. Girls with formal qualifications and girls who said they liked the 'interest' of their work reacted against office automation, the former because, like the bank girls, they saw their qualifications becoming irrelevant, the latter because they liked the *status quo*. In contrast, the Hollerith girls, with their knowledge and experience of mechanisation, were more likely than other clerks to look favourably on the idea of a new work system. We had not been able to identify any relationship between attitudes to change and class background or education.

In the post-change replies there was again no relationship between class background or education and attitudes to automation. Also there was little consistency in the pre-and post-change attitudes of the seventeen clerks whom we had been able to interview twice. Those who had approved of the 'idea' of automation did not necessarily like the 'reality' and vice versa. In fact, only two clerks gave the same replies in both interviews. Attitudes were therefore not 'set' in any way before the change. It seemed that individual clerks had changed their views according to how the computer had affected them personally and how they anticipated it would affect clerks in the future.

However, one fact that had influenced pre-change attitudes still seemed to exert an influence. Girls who said they liked the 'interest' of their work had been against the 'idea' of office automation and, after the change, we found that girls who thought their work was interesting were more likely than other clerks to say the computer had no advantages for the firm. None of these girls worked on the new computer jobs. It seemed that if clerks found their work interesting and enjoyable, they were more likely to be resistant to both the idea and the actuality of change; they did not want a pleasant situation to alter.

Girls with formal qualifications, a group which, in the pre-change study, had suggested automation would prove damaging for clerks, had swung round in their opinions. They were now more likely to say that the computer had brought them personal advantages.[4] This group consisted of typists, comptometer

[4] This is not the same group as in the pre-change part of the study. Some girls with formal qualifications had left and others had joined the firm.

operators and general clerks, who had noticed a lessening of their work load after the introduction of the new system, but who had suffered no particular disadvantages. Apparently they were now less pessimistic that the computer would make their qualifications redundant.

Although shared characteristics such as liking job 'interest' and having formal qualifications made clerks tend to react in the same way, membership of a common work group seemed one of the most important influences on attitudes. In the previous section we noted how the Hollerith group had approved of the idea of office automation and how the newly formed computer group and the remaining Hollerith personnel still approved after the change had taken place. We found that the computer and Hollerith groups were more likely than other clerks to express approval of the changes that had taken place in the office; they were more likely to say that the computer had no disadvantages for the firm, and computer and Hollerith machine operators were the only clerks suggesting that the introduction of computers into offices had improved the clerical career prospects of girls. It seemed that this group had begun to realise that the firm's dependence on the new work system made their skills indispensable. One told us 'If I decided to leave, the firm would be in trouble, no-one else here could do my work'.

Our evidence from Carters seems to support the hypothesis we put forward after analysing Royal Exchange data—that an important influence on attitudes to change is how individuals see a change as affecting themselves and their personal interests and ambitions. It also demonstrates that these attitudes can change as experience of an innovation shows pre-conceived ideas to be erroneous, and that group membership and shared experience and beliefs will exert a considerable influence. At Carters the existence of an already mechanised work system meant that girls were used to and liked machines, and those girls still occupied with traditional clerical tasks looked on machine work as promotion. The fact that the clerical group as a whole was not highly educated was an advantage from the point of view of the firm, for routine, mechanical tasks were quite acceptable to them, whereas they were not to the Royal Exchange bank girls.

Although we could find no direct evidence that class background influenced attitudes to change, it seems likely that the first hypothesis was at least partially valid and that the girls' working-class background made machine work a perfectly good way of earning a living, whereas for the middle-class bank girls it had unpleasant associations with low status factory work.

The handling of the change process

We concluded our analysis of Royal Exchange data by examining the strategies employed by management to introduce and carry through the change, and the clerks' reactions to these. In Carters it was difficult to do this. Although the Carters programmers had had so many problems in getting the computer operational, none of the clerks seemed aware of these. It is likely that the existence of the Hollerith installation, and the fact that work could always be processed on this, lessened the impact of technical difficulties on staff outside the computer centre.

Problems in the Accounts Department seemed of a very minor nature. A few punch operators referred to the difficulty of learning to use new machines and there were some complaints about the heavy work load involved in changing from one system to another, but there was little else. However, it will be recalled that the firm took three years to transfer part of its accountancy procedures to the computer, and so there was plenty of time for adjustment. This, together with the existence of a mechanised system before the computer arrived, meant that the traumas of change were far less for Carters' clerks than for the bank staff. In Royal Exchange, although the change-over to E.D.P. had been even more drawn out, individual branches had felt the impact of the new system more sharply.

Although we have criticised Carters for its secrecy policy, in fact not one clerk did so, and there were no suggestions that management should have kept the clerks more adequately informed of what was to happen. The Carters girls did not expect to be told anything, and there were no complaints when they were not. Managers might interpret this reaction as showing that talk of the benefits of good communication is so much eyewash. Being too frank is only going to stir up trouble that might otherwise be avoided. We suggest they should however, be cautious in their deductions. We shall show in Chapter 10 that communication needs differ, depending on many other variables in the situation. Carters' clerks did not feel threatened by office automation, therefore they were not very anxious about its effects. They had no vested interest in or commitment to their jobs. It would be easy to find others. Lastly, they had low expectations of management. It had never communicated or consulted with them before, why should it now?

PART III
The analysis of change

CHAPTER 9

Change and its consequences

In this chapter we will re-examine our case-study data and attempt to draw some conclusions on how automation of the clerical function may affect an organisation and its employees. The consequences of a change-over to electronic data processing will be considered first from the point of view of the firm and its policies. Next we will see how the change affects employees and their jobs.

The Consequences of change for the organisation

The consequences of change will always depend on the nature of the innovation and the extent of its impact on the organisation. Carters had adopted the 'hardware' rather than the 'systems' approach to computers; that is, it had merely transferred routine clerical procedures on to the machine. The bank had made some systems changes but there was no extensive re-organisation of the clerical function as a whole. In addition, both firms invested in medium sized computers which could only handle a restricted amount of data processing and this in itself limited the impact of automation on the firms' organisation. American research suggests that the changes brought about by small or medium sized computers will differ qualitatively from those initiated by the installation of a large computer (Mann 1962). Our research has produced no evidence that would refute this hypothesis.

In neither of our firms did the stage of automation we observed lead to any major organisational changes, but it did seem that the prior existence of a punched card installation in Carters caused the organisational impact of the computer to be slighter than in the bank. There was very little alteration in the content or skill of work and, in the computer area at least, staff went on doing the same kind of job they had always been accustomed to.[1] It must be stressed, however, that although, so far, there has been little organisational change, if Royal Exchange and Carters continue with their automation programmes, and the bank certainly intends to do this, it is likely that there will be a move from a 'hardware' to a 'systems' approach and considerable organisational re-arrangement will then result.[2] At the end of this section we will indicate what kinds of changes might be involved.

All innovations have their positive and negative sides; they bring with them both advantages and problems and, if the change is to be worth while, then the final gains must outweigh the problems. In Royal Exchange, clerks and managers believed that from the bank's point of view this had happened. The computer had reduced the need for scarce labour in a time of expanding business. It had speeded up work, increased efficiency and maintained the bank's status in relation to its competitors. Even more important, the move to automation had brought with it a tremendous potential for increasing efficiency and reducing costs in the future. In Carters, the technical unreliability of the first machine left a question mark over whether advantages had surpassed disadvantages. Again, benefits would seem to lie in the future. The firm had obtained valuable experience of automation problems and had built up a staff who were experienced in computer operation.

Neither firm appeared to have gained all the possible advantages from their computers, and this was almost certainly a result of their pioneering role and the lack of information about E.D.P. available in this country at the time of their investment. For example, American studies suggest that, apart from advantageous changes resulting from the computer itself, the machine can bring about a psychological 'unfreezing' of a situation and provide top management with the opportunity for reviewing their organisations and introducing long term changes (Mann and Williams 1960).

[1] Experience with a punched card installation did not remove all Carters' problems for they still had considerable trouble in making the systems change. This rather conflicts with American research findings. Mann says that 'installations of this type take only six months and are felt by employees to be only mildly disruptive.'

[2] Most of the Big Five banks already pick up data automatically from cheques and feed it straight into the computer. This transfers a major part of data preparation from the branch to the computer centre.

E.D.P. thus provides the opportunity for comprehensive change irrespective of its own functions. This was not appreciated by Carters, and the bank preferred to proceed slowly and carefully.[3] Both firms were very cautious in their approach to the new system and content with very limited gains.[4] But in view of the problems that were encountered, it can be argued that they were wise to be conservative. Carters would not have had the know-how to introduce any more fundamental changes at that stage, and the bank would not have wished to do so.[5] Both organisations had difficulties in assimilating change even at the slow pace it was introduced. More accelerated change would have thrown an intolerable strain on firms which were geared to stability or a slow rate of change.

Let us now turn our attention to some of the organisational problems of change and see with which of these our firms had difficulty. The most serious and unanticipated problems seem to have emerged from the change process and the handling of this rather than from the consequences of change. In both Royal Exchange and Carters, management and computer staff saw the installation of a computer as nothing more than a change of system and did not anticipate its social and organisational effects. The problems of the machines' introduction and assimilation were increased by the unexpected length of time the changeover took and this in itself created extra psychological and, for Carters, financial strains.[6]

[3] Stafford Beer, writing in the *Manager*, October 1963, says 'never ask how to automate something that is there, always ask whether automation can be used to procure a better solution to the management problem that is currently solved by what is there.'

[4] It will be remembered that the computer manufacturer who supplied the bank machine complained that management were so wedded to traditional banking procedures that they wished to retain in the new system all the checks incorporated in the old. This slowed up data processing.

[5] Mann and Williams (1958) point out that 'organisations vary markedly in their capacity for change. Resources—such as previous experiences with change, managerial talents, work force skills, values and attitudes —are important conditioning variables'.

[6] Greensmith (1959) says 'widely differing figures are quoted when one talks with companies as to the time taken in changing over to E.D.P. To some extent the time taken depends upon the degree to which the computer and ancillary equipment have been proved in practice'.

The stress which developed in both computer centres was to a large extent due to management concern over the delays and breakdowns and to their pressure for faster results at a time when these were not attainable. In Royal Exchange, the Ringer Square branch, in particular, felt that after many months of waiting the new procedures were implemented with excessive speed. Particular aspects of the firms' change policies, such as communication, consultation and timing, will be discussed in some detail in the following chapter as they proved to be important conditioning factors in the clerks' reactions to the computer.

Both Royal Exchange and Carters encountered the kinds of personnel problems which any firm installing a computer must expect. For example, they both had difficulties with training. In Royal Exchange the worst of these was associated with the new computer jobs. Bank clerks had to be trained from scratch as programmers and computer operators, while young girls straight from school were recruited to look after the ancillary equipment. The programmers were sent on a two weeks course provided by the computer manufacturer and then left to pick up the techniques for themselves, with assistance from the computer manufacturer's representatives. For many months they were inexperienced and unsure of themselves and this uncertainty coincided with the period when the computer was technically most unreliable. This raises the question of whether a firm should rely entirely on newly trained staff for running a first computer. The manufacturers of Royal Exchange and Carters' computers both suggested this but it does leave a firm in a vulnerable position—entirely dependent on the computer manufacturer's good will. At least one qualified computer technologist would seem a necessary insurance policy. Unfortunately when our two firms were installing their computers these specialists did not exist.

Royal Exchange also had problems in training operators for the computer ancillary equipment and the branch punch machines. The supervisor responsible for girls in the computer centre was used to the old kind of bank mechanisation involving ledger posting machines, but not to the new. She had to learn new procedures and techniques herself while at the same time attempting to teach these to the girls working in the Centre.[7] Similarly, in

[7] Kushner (1963) suggests that training can be one of a firm's biggest problems during the conversion period. At the moment it seems that

the branches, girls accustomed to operating ledger posting machines had to be transferred to the new punch machines. Although some were sent on a manufacturer's course, others told us they had had to pick the job up for themselves. Lack of expertise combined with technical difficulties caused a great deal of faulty input tape to be produced in the early months of computer operation.

In Carters, training problems in the computer centre were fewer because the girls were already experienced punched card operators and punchers. Even so, the transition from one kind of machine to another proved more difficult than had been anticipated. In this firm the greatest problems were encountered in departments where clerks had to be more accurate than previously and had to learn new forms of code. Yet the clerks were not anxious to co-operate in learning these techniques for they saw the computer as a threat rather than an advantage.

Training problems then caused additional difficulties at a time when staff were trying to overcome the organisational and technical problems arising out of the new system. Some anticipation of the likelihood of this kind of problem arising and the introduction of formal training schemes run by experts would certainly have eased the situation.

Another problem encountered in our two firms, and in all the others we visited during the course of this research, was the tremendous amount of work involved in changing over from an old system to a new. Staff in both Royal Exchange and Carters complained about the vastly increased work load they had to carry during the change-over period and this was as true of staff in affected departments as in the computer centres. Computer centre staff, because they were highly identified with the new system, were at first very willing to accept this as an unavoidable part of their new responsibilities but, as difficulties and pressures increased, in Carters particularly, their morale cracked and their health started to deteriorate. In affected departments staff were against, rather than for, the computer and here the work of preparing for an E.D.P. system caused a lot of grumbling and some discontent. The clerical staff of neither firm had been prepared for so much work, although the bank was in a more fortunate position than Carters in

many firms are training computer staff for their competitors. One local retail house has lost five programmers—all lured away by promises of higher rewards.

that it could transfer clerks for a temporary period to help the hard pressed branches.

Consideration of the increased work demands of change leads on to the question of financial rewards and this too will be discussed in the next chapter. Mann and Neff (1961) point out that 'while it is generally believed that the incentive of greatest importance is economic—it is not uncommon to find management expecting the key men responsible for directing the change, and even the non-supervisory employees, to work harder than they have ever worked before and get only the same, or even less pay than usual'. Certainly the programmers in both Royal Exchange and Carters were dissatisfied with their pay and believed they were not being adequately compensated for the additional work and overtime. Similarly, a number of senior branch management expressed the hope that the bank would provide some material recognition through the Christmas bonus scheme. The knowledge that there would be some financial compensation for additional work and for the psychological and physical stress that accompanied it would have helped to maintain morale in both Royal Exchange and Carters. It can be argued that some reward for the hard work that goes with this kind of systems change is likely to act as a stimulus and will help to induce favourable attitudes in those involved in the innovation.

Surprisingly, our initial belief that staff would be displaced by a computer and that this would result in either redundancy or the transfer of a large number of clerks to other departments was not fulfilled. Carters' clerks and some of the bank clerks were worried about the possibility of this and, logically, if a computer takes over a manual clerical operation and is used effectively, it should cut down staff. Computer manufacturers make staff saving an important sales point. Firms introducing computers must therefore, if they are far sighted, make advance plans and policies for handling staff displacement.[8] Neither of our firms did this, nor as it turned out did they need to, for neither had any redundancy nor even any

[8] Mumford (1964) describes how, if plans are made early enough, a labour force can be run down to computer requirements through natural wastage, although temporary staff may have to be engaged to cope with the heavy work load of the parallel running period. If surplus staff can be absorbed elsewhere in the firm then suitable jobs and departments must be identified well in advance.

serious labour displacement. Carters ran its old and new systems side by side until it abandoned its first computer, and the bank was both short of staff and increasing its business; it therefore had a built-in mechanism for dealing with labour displacement and could provide reassurances about redundancy. It seems certain, however, that less clerical staff will be required in both firms as more functions are transferred to the computer, and they will eventually have to formulate policies for running down their labour forces and re-assigning staff displaced from automated departments. Usually it is easy to achieve a rapid reduction in the numbers of young female clerks, as this group has a high labour turnover anyway.[9] However, redundancy policy is a matter that trade unions are actively interested in and Royal Exchange may encounter difficulties here because so many of its employees belong to a union, N.U.B.E., that it does not recognise.

Although, up to date, the organisational impact of office automation has been slight in our firms, if they continue to transfer work onto their computers, then more fundamental changes must take place. Once the present 'hardware' approach is rejected in favour of integrated data processing—or a 'systems approach'—then quite radical changes in management and departmental structure are likely to occur. As any firm with a computer is wasting expensive equipment if it merely uses the machine to complete existing functions more rapidly, it seems reasonable to predict that our firms will eventually move towards integrated data processing. Let us briefly see what organisational changes will have to be faced then.

Once a firm uses its computer to provide a centralised data processing service for the control of the total company operation, then change will occur in its information system, its decision making procedures, its departmental organisation and its management structure.[10] There is no longer any need to employ clerks to

[9] The problem, from a community point of view, may not be redundancy so much as a reduction in the number of clerical jobs available for school leavers. In an area where there is a teenage unemployment problem there could be serious employment difficulties if a large number of firms cut down their labour requirements at the same time. Walter Buckingham, Director of the School of Industrial Management, Georgia Institute of Technology, refers to this process as 'silent firing'.

[10] Mumford and Ward (1965) provide a more detailed discussion of the impact of computers on management organisation.

prepare statistical data and feed it upwards. Providing senior management decides in advance precisely what it needs to know to run the business, information can be sent to it direct from the computer centre. This in turn affects decision making processes. Senior management is now in possession of the kind of information which enables it to make policy decisions without referring to the lower ranks of management. The computer will have completed the preliminary analysis of variables which previously middle management would have been responsible for. The computer can therefore remove some of the decision making functions of middle management.

Once the new system has altered information and decision making processes then this, in turn, will affect the departmental structure of the firm. If the problems underlying traditional departmental procedures are handled electronically then there is no longer any need to retain departments in their old form. And so clerical departments handling records, or compiling statistical data, or exercising control functions may be eliminated, combined or drastically reduced in size. For example, the conventional wages department will be inappropriate once computers compile earnings and issue credit notes which go straight to an employee's bank.

All these changes affect a firm's management structure. One consequence is a flattening out of the management hierarchy pyramid now typical of most firms, for the preparation of so much data in the computer centre can lead to a reduced need for middle management. It can also widen the span of control possible for senior management as continuous interaction with subordinates becomes less essential. The classical theory that a manager cannot control more than a limited number of subordinates was based on the difficulties of interacting with a large group of people.

Lastly, changes in organisation and structure lead to change in a firm's power structure. Spencer and Sofer (1964) point out that a realignment of priorities by an organisation must give greater power to some departments at the expense of others. With an extensive use of computers one would expect to find a reduction in the power of middle management, an increased restriction of power to the top of the enterprise, some realignment of departmental power and the emergence of a new and powerful group in the computer technologists. In Royal Exchange there was already some indication that a shift in power was taking place. Whereas, at Carters, the computer centre was largely incorporated into the Accounts Department and supervised by the senior accounts clerk, so that the old structure was retained, in the bank, branch autonomy of operation was removed by the computer. The branches were now left with only part of the accounts procedure to process, and were dependent on the computer centre for accounts information. Morenco (1965), studying the introduction of a computer into a French bank, points out that whereas senior management of the bank saw the computer as an instrument for providing branch management with more accurate information on which to base decisions, the branch managers did not welcome this and in fact reacted to the computer with hostility. She believes this hostility was 'due to fear of a threat to the strategic position of a sector which hitherto had been shielded by its specialist character from all intervention'. Although we have no clear evidence on this point, it would seem likely that the British bank managers reacted in the same way, for the same reasons, and that many of their objections to the removal of accounts data to the computer centre were because they believed this would lead to a weakening of their power position.[11, 12]

Besides altering the power relations of departments, the introduction of an automated system also gives power to new groups or to groups who have not had it before. In both Royal Exchange and Carters, programmers were now of tremendous importance; neither firm had many of them and, in view of their scarcity, could not afford to lose those it had. Yet Carters paid its first programmers very badly and the bank gave them only a small increment above its normal salary scale. The unprecedented departure to outside industry of bank clerks who had become

[11] Siegman and Karsh (1962) found that when a computer was introduced into a civil service bureau, 'one of the most notable effects was the mounting power and authority of the specialists who directed the E.D.P. operations. Hitherto relatively independent functions became totally dependent upon their operations and directions.'

[12] Rearguard actions are sometimes fought to prevent this shift of power. The research team heard of one firm which had three computers, each one controlled by a different department. Proposals for an integrated data system were successfully blocked by managers who wished to retain absolute control of their own functions.

programmers caused this policy to be revised.[13] Computer engineers are also essential members of staff and, from a firm's point of view, more of a risk than programmers, as usually they are on loan from the computer manufacturer. As this is being written the engineers of one computer manufacturer are on strike and causing considerable embarassment to the firms using their services.

The bargaining power of these computer personnel is, of course, partly a result of their scarcity value. As more programmers are trained and computers are autocoded for standard commercial procedures, the market should become more stable. However, this is not likely to happen for some years to come. Experienced computer users anticipate a serious shortage of programmers over the next few years. This will hit particularly those firms who wish to use their machines in a sophisticated way and therefore need highly skilled computer staff.

Unsophisticated groups such as punch girls will also have power once an organisation has much of its work on a computer. A situation can be imagined where a strike of punch girls could paralyse the office work of a firm. Usually, of course, groups like this are not unionised and are unaware of their power potential. Unless very badly used by a firm they will not be inclined to exploit their position.

These then are some of the organisational problems of change that Royal Exchange and Carters are likely to meet with if they extend their computer usage. The more complex an innovation is or becomes, the more complex become the problems of change and adjustment which it involves.

The consequences of change for the group and the individual

Some consequences of using computers, such as the need for shift work, the increased importance of meeting work deadlines, the difficulty of both finding and concealing errors, and the new

[13] Managers frequently complain that computer specialists have developed a jargon of their own which inhibits communication. Some jargon inevitably develops with a new technology, but it can be suggested that when this is used to excess it may be an unconscious attempt on the part of programmers to give their work a 'mystique' and thus reinforce their power position.

machine environment, will affect many groups of employees in a firm; other consequences will affect only one or two groups or will affect a number of groups in quite different ways. In this section we will look at the main occupational groups in Royal Exchange and Carters, and examine the implications of the new E.D.P. systems for each of them in turn.

PROGRAMMERS

In Royal Exchange, in particular, the programmers were a most interesting group. Bank clerks brought up in traditional banking methods and attitudes were now allotted completely new roles—those of innovator and technical specialist. Instead of spending their days in clerical routine and established practice, they were expected to transform themselves into 'ideas men', capable of working out complicated systems for an entirely new work medium—electronic data processing—and able to put these systems into practice and to get them accepted and adopted by bank staff in other parts of the organisation. Recruited from the bank's Head Office and from branch banking, they were now isolated in a new computer centre which was physically separated from all other sections of the bank.[14]

Once the programmers had adopted their new roles, besides being in a new, exhilarating and demanding work situation, they appeared to undergo a change in group goals. Instead of wishing to maintain work stability and preserve the status quo, they came to have a vested interest in change, for both the importance and the interest of their work lay in designing and trying out programs for the new system.[15] Once this stage was completed and a new work procedure established and implemented, the excitement and challenge disappeared. On this point March and Simon (1958) comment 'it is frequently observed that the initial stage of program elaboration is generally a period of excitement for the personnel engaged in it. They put in a great deal of overtime, and take much pride and pleasure in their work. As programmed

[14] March and Simon (1958) suggest that 'the creation of a new unit is the only way to secure innovation that is not excessively bound and hampered by tradition and precedent.'

[15] Crozier (1964) provides an example of how managers in a highly bureaucratic company could only increase their power by introducing change.

activity begins to replace innovation, excitement wanes and feelings of anticlimax are often expressed.'

A group with an entirely new approach to work had therefore appeared in the bank environment. Their work interest depended on change, they had no contact with customers and little contact with branch activity; in addition they had considerable market value and might easily be tempted to leave the bank for better paid jobs in outside industry. The bank, for the first time, was presented with a junior management group who were no longer completely identified with banking. It will be interesting to see how the bank establishment reacts to this new and alien computer group. In industry it is quite usual to find firms very wary of their non-conformist specialist staff and inclined to regard them as potential sources of organisational infection. Kahn and Bolding (1964) point out that this leads to 'maladaptive segmentation'. The specialist becomes absorbed in his personal career goals and is not involved in the goals of the organisation. The firm, in turn, sites its laboratories or its computer centre in buildings a mile or two down the road from the main factory so that its technical specialists are kept as far away as possible from other employees.[16]

Signs of a gulf between the computer specialists and branch management were already apparent in Royal Exchange and led to complaints by branch managers that the programmers did not take enough account of branch problems and spoke a technical jargon that branch staff could not understand. As these programmers had been bank clerks themselves only twelve months previously, it was organisationally retrogressive that they had adopted their new technical role so rapidly and apparently forgotten the realities of branch banking.

Many of the points made above are true of Carters programmers, although, as the firm recruited them especially for the computer centre, they did not have such a major change of role. Also, the fact that clerical procedures were already mechanised in Carters meant that the computer's impact on work environmnet was much less. Nevertheless, the fact that Carters did not appreciate the importance of its new specialists meant that it nearly lost both of them in he early stages of the computer's installation.

Croome (1960) also discusses this phenomenon.

MANAGEMENT

The management group can be considered from three different aspects. First, the effect electronic data processing has upon the nature of the management job; second, the kind of reactions that may be expected from managers presented with the new system; third, the implications of our research and other comparable research, for senior, middle and junior management.

The effects of a computer on the power structure of a company have been discussed earlier in this chapter and it has been suggested that middle management, at least, is more likely to lose than to gain power through the use of E.D.P. Increased power will be found in the new specialist groups and at the top of the organisation, where it is now easier to locate all major decision taking. We have suggested that this shift in power did occur to some extent in both our firms. We have also referred briefly to changes in decision making functions and it is worth considering these now in terms of the manager and his job. Whisler and Schultz (1962) point out that the use of a computer places strong emphasis upon rationalising and qualifying decisions; many of these decisions, or as they may now be more appropriately called, logical deductions, are then incorporated into the machine, leaving managers to spend an increasing percentage of their time on problems involving interpersonal relationships. Where a manager retains his decision making functions—and a computer can never remove them entirely from him—these may be changed and simplified. For example, the number of variables he has to consider when taking a decision may be reduced. Our research did not show up any major changes in the decision taking functions of Royal Exchange or Carters managers, but Morenco (1965) in her French Bank study describes how managers responsible for providing customers with discount facilities were given a risk analysis by the computer and were left with only the final decision on whether or not to allow a claim. Previously they had had to weigh up the various factors affecting a claim and arrive at a decision on the basis of their analysis. One can predict that this sort of change will occur in Royal Exchange as the change-over to E.D.P. progresses. Although Whisler and Shultz suggest that the removal of decision making functions will free the manager to devote more time to personal relationships, it must be pointed out that many managers believe decision making to be the most important and enjoyable part of their job and they will not lightly relinquish this in order to assume a role which many have not been trained to see as of major importance. Also, if the numbers of clerical staff are reduced through automation, a manager may not have all that many staff to look after.

Argyris (1962) makes another interesting point about managers and decision taking. He found that with traditional procedures, relatively incompetent executives were able to hide their limitations because they rationalised that 'no-one could possibly have enough of the relevant facts to make a decision'. With the use of a computer, the number of facts that they could consider was greatly increased and so was their anxiety about making decisions. Bad decision taking would be less easily excused by senior management because crucial variables could be identified by the computer. Kushner (1963) also takes up this point and suggests that because a computer can show up inadequate managers, deficiencies of past policies of managerial recruitment and promotion are also apparent.

One other illustration of computer impact on the managerial job is of interest. This is the removal from middle management of time for manoeuvring, covering up, or ideas selling. Argyris (1962) cites one firm where, before a computer was installed, it took 3-4 weeks for the President to receive feedback about the way certain decisions were implemented. This allowed time for managers to rally support and to set out their unpleasant information in such a way as to soften the blow. With the advent of the computer, it became possible for the President to receive necessary information in two or three days. His reaction was, 'I'll now be able to nail things down and find the real culprits.' We did not come across any uncovering of incompetent executives or removal of 'political' time from management in our case study firms, but this kind of thing could well happen in the future.

Kushner (1963) has categorised managerial reactions to computers in the following way. First, there is the manager who sees the computer as an opportunity for progress, either for himself personally or for his firm. The two managers in Royal Exchange and Carters who persuaded their firms to order computers would come into this category. Second, there is the manager who sees it as a routine change, is neutral about it, and pays little attention to the new developments. No affected managers came into this category at either Royal Exchange or Carters, although non-affected managers at North Office did react in this way. When it is found in affected managers, this reaction might be an unwillingness to face unpleasant facts.[17] Third, there is the manager who is anxious about the new developments and worried about his own status or job—most Royal Exchange and Carters managers affected by E.D.P. had these fears. Fourth and last, there is the knowledgeable manager who recognises that a computer will have both advantages and disadvantages and is prepared to see what use it can be to him. The crucial word here is 'knowledgable'; this kind of evaluation cannot be done without a complete understanding of the consequences and potential of the new system, which our two case study groups of managers did not possess.

The way managers react to a computer will depend not only on how they see it as affecting themselves and their jobs, but also on their own personalities and their capacity for adaptation to change. Selznick (1957) points out 'that individuals whose ways of thinking and responding served the organisation well in an early stage, may be ill-fitted for the new tasks.' He suggests that this is not so much a matter of technical knowledge as of attitudes and habits. We have seen how the managers of Royal Exchange, brought up in an atmosphere of banking tradition and slow change, were unenthusiastic about the computer and its consequences. Although their hostility seemed primarily due to fears about their personal status and the effect of the computer on their job, it can be argued that some of it was also due to the fact that people who welcome a great deal of change do not normally choose banking as a career, and that it may be the most cautious, although perhaps also the most responsible, bank clerks—the two may be synonymous in banking—who are selected as managers. In neither Royal Exchange nor Carters would one expect to find the very dynamic, ambitious type of young manager who is making rapid progress upwards and welcomes change as a challenge and a stimulus.

Attitudes and habit were therefore likely to be elements influencing managerial reactions in Royal Exchange and Carters, although a lack of technical knowledge also seemed to be an important additional factor. One manager, for example, told a

[17] The research team met one O. & M. manager who had tried to consult and communicate with managers about to be affected by a computer. He found that they were quite uninterested and apathetic.

member of the research team that 'in the old days if there were any errors I would go out into the office and help my staff trace them. I could do this because my twenty-five years business experience told me how most errors were likely to occur. Now we have the computer, if there are errors I keep out of the way because I don't know enough about the new system to be any assistance.' This difficulty the average manager has in understanding E.D.P. seems an important problem that needs to be solved quickly. Men who are in charge of departments must know what is happening if they are to retain administrative control. Some means must therefore be devised for giving them a rapid yet thorough understanding of computer procedures.[18] This raises problems of management training to which few firms are, as yet, paying much attention. Hart (1962) the principal of a British computer training school, has asked the question 'Can our present methods of management teaching be said to prepare our managers for the sort of business world which computers could give us in fifteen years or so?' His answer is 'No'.

What general conclusions for management can be drawn from the points made above? To start at the top, it seems of the utmost importance that senior management, when they take a decision to invest in a computer, should themselves be aware of the implications of the step they are taking. They need to realise that they are not merely buying a piece of equipment, but are starting off a chain of reactions that will affect their firm's organisational structure and the jobs, responsibilities and status of managerial and clerical employees. They should appreciate that the change will involve major policy decisions and must be handled in such a way as to make it acceptable to staff. In order to formulate these policy decisions and handle the change, senior management will need to analyse the firm's total organisational and social environment, identify how this is going to be altered by the new system and predict the effects of the change on the firm's staff and structure. In our firms, one reason for the inadequate handling of the change

[18] A systems expert who had installed many computers told the research team that he had great difficulties in getting older managers to understand how a computer would affect the work of their departments. As he talked to them he could see them gradually losing contact. At last in desperation, they would produce one of their bright young assistants and suggest the explanation should be given to him.

process was the fact that this was introduced by an O. & M. specialist in Royal Exchange, and an accountant in Carters. Quite naturally these managers looked at the change in the light of their own particular specialisms and were unaware of the important social implications of the new system. No preliminary analysis to ascertain the effects of the computer installation on the total situation was undertaken and many computer consequences, such as the anxiety shown by branch managers, were not sufficiently catered for.[19] The dilemma for top management is how to further organisational goals while at the same time looking after staff needs. It can do neither of these without preliminary analysis, prediction and policy formulation.

Middle management, particularly the managers of departments whose work is being transferred to a computer, are presented with particular difficulties during the change period. It is hard to understand why Royal Exchange and Carters did not involve these managers more in planning the changeover. They had much to contribute through their intimate knowledge of existing work organisation and their influence over their staff. Effective human relations are important during the early stages of change when staff are feeling uncertain about what is going to happen, and during the later stages when excessive work and overtime is the norm. Middle managers have a key role in translating the organisational objectives and policy to subordinates in terms that they can understand and accept.

CLERKS

The group that we studied in greatest detail in our research was, of course, the clerks. Let us now examine how the jobs of clerks in both Royal Exchange and Carters were altered by the computer, and compare our conclusions with the findings of similar research investigations in other countries.

Environment In Chapter 2 we suggested that the division between manual and non-manual work is breaking down and that the development and extension of mechanisation brings a factory type environment to the office. Office automation implies the extensive use of machines—and therefore noise—the introduction of shift

[19] Mann and Neff (1961) suggest that 'a major crisis in the planning and execution of a changeover may be necessary to ensure real understanding of the change by key top level personnel.'

work, and a loss of work freedom through the increased necessity of meeting deadlines. Some of these environmental changes were accelerated in our firms as a result of the computer but there was no extreme alteration in physical conditions. One reason for this was the fact that Carters and Royal Exchange had already, to some extent, moved away from the traditional office situation, for both had machines before the arrival of the computer. Carters operated the equipment associated with a punched card installation, while the bank used adding and accounting machines. The replacement of these by punch machines for the computer was not therefore of much moment and caused little deterioration in working conditions

Both firms did, in fact, introduce shift working to some extent but in Carters this was confined to programmers and computer operators, and in the bank to ancillary staff located in the computer centre. Although overtime in the branches was usual during the transition period, once this stage was completed bank clerks found that, if anything, they could get away earlier in the evening than before.

The need to meet deadlines did become stricter in both situations, but here again neither of the two firms had ever had a great deal of freedom in the planning and completion of work. Carters had enjoyed rather more flexibility in this respect than the bank where all transactions must be ledger posted and balanced the same day.

In our two case studies, then, the computer's impact on the normal office environment was slight, although the machines themselves were housed in rather special surroundings. Carters made do with a converted ground floor office, but Royal Exchange had a specially designed computer centre with modern decor, air conditioning and temperature control. But, although branch working conditions were little affected by the new system, for some clerks work was considerably altered and for a large number it was altered in some degree. On the whole, Carters' clerks experienced slight change and approved of the new system, but Royal Exchange clerks, who were more affected, found few personal advantages and there was evidence of a reduction in job satisfaction.

Job grading and promotion A number of writers have suggested that electronic data processing leads to an increasingly specialised division of labour, to a reduction in the number of different job grades through the elimination of certain activities, to a polarisation of jobs towards the higher and lower end of the skill ladder and to changes in job content.[20] Hoos (1960), after studying twenty American firms, claims that 'dismissals, downgrading and transfers' are frequent immediate results. Again, none of these effects were very noticeable in our firms. Machinists and control clerks in the bank were left with rather narrower functions than before and the numbers of clerks required to complete specific tasks were reduced, but no jobs were entirely eliminated: on the contrary, some new jobs were created in the bank—branch computer controller for example. This meant that we did not observe any polarisation of work although there was some indication, again in the bank, that low status jobs, such as machining, were even further depressed in status by E.D.P. We came across no instances of downgrading and encountered few transfers. However, this is not to suggest that this kind of thing does not happen. The small amount of job and organisational change resulting from our two computers was due entirely to the very limited work areas in which they were operating. As automation proceeds in the two firms, so the impact of the computers will become more noticeable.

Although neither Carters nor the bank had to downgrade staff, the bank at least was able to do some upgrading by transferring clerks from routine banking to the more skilled jobs in the computer centre. In the branches, a small number of clerks were made computer officers responsible for liaison with the computer centre but, apart from this, branch job grading remained unchanged. This differs from the experience of some American firms but accords with that of others. One study of an American insurance company (Ginsberg, 1963) found that there was considerable upgrading of the labour force after the installation of the computer. In contrast, Mann and Williams (1960) report on a light and power company where nearly everyone anticipated higher job grades but these did not materialise. The most menial jobs were eliminated but so were a number of high level non-supervisory jobs. However, it is impossible to generalise on the amount of effect a computer will have in this area. The significant variables are always the nature of the functions which the machine takes over and the extent to which work is transferred. Broadly,

[20] See, for example, Kushner (1963).

it can be said that if jobs such as record keeping, filing, computing or tabulating, keypunch and related machine operations are automated, then, if the computer is being used properly, a considerable number, although not all, of routine jobs will be eliminated.

We had anticipated that clerks would see the introduction of an automated office system as providing them with new promotion opportunities. In fact the opposite happened. Male bank clerks who before the change had believed their promotion prospects to be good, after the change were more pessimistic. This again accords with American experience. Kushner (1963) reports that in a Public Utility company employees saw less opportunity for promotion after the change than before. He points out that this was a realistic conclusion since many senior or technical clerical jobs had been eliminated. We did not, of course, encounter this elimination of jobs in our firms but again it is a logical outcome of any extensive use of electronic data processing. Our male clerks may have suspected this fact although they did not know enough about the consequences of automation to be certain.

Job status Job status is another important area of job satisfaction, and research has shown that the one tends to correlate with the other. In Carters and Royal Exchange we had the interesting difference that in one firm job status was thought to be enhanced through the computer, in the other it was seen as depressed. In Carters, machine jobs had a high prestige before the change and an even higher one afterwards. But in Royal Exchange, before and even more after the change, machining was the job most disliked by the girls, and the majority of girls performing it wanted to move to something else. This contrast in attitude was due partly to a difference in background and aspirations. Carters girls were working class with secondary modern school education; any office work was a step up for them and machines did not have undesirable associations with factory work. Most of the Royal Exchange girls, on the other hand, were from middle-class families and had grammar school educations and G.C.E.

Another important factor conditioning attitudes was the difference in role of the two groups. In Carters, girls on machine jobs worked either in or near the computer centre and were part of the new specialist group responsible for introducing the new system. They therefore had an innovating role. In Royal Exchange, the girls had nothing to do with the computer centre and most had never seen it. They were on the receiving end of change—their accustomed way of life was being disturbed by a remote and mysterious group of specialists located in another building. The distinction between being on the implementing as opposed to the receiving end of change seems an important one, for it separates out the 'agents' from the 'victims' of change. Similar reactions were found in the French bank study where punch operators attached to the computer centres thought their jobs now had prestige and status and that they had become more important to the bank. Informal conversation with girls operating machines in the Royal Exchange computer centre led to a strong impression that they liked being associated with a new and modern development in banking, that they found their work status conferring and that they were more satisfied with it than girls on similar jobs in the branches.

A third factor conditioning attitudes must, of course, be the effect automation has on job content. This will be analysed in some detail in the next section. The point can be made again here that Royal Exchange machinists and control clerks did find that the new system had narrowed down their jobs and made them more routine. This did not happen in Carters where the changeover from a punched card installation meant that the content of machine jobs remained virtually unaltered.

Job content It seems clear that when a firm moves to a computer from a punched card installation clerical work is not greatly changed. In three American firms changing from a mechanised system to automation, a research team from Michigan State University used the fourteen questions on different job aspects that we also used in Carters and Royal Exchange.[21, 22] The common factor in all three firms and in Carters was the large number of clerks saying that the computer had not greatly affected their jobs. If we refer to the American firms as A, B and C respectively, in American firms A and C, two-thirds of all employees said their jobs had not

[21] For details of the American research findings see Hardin (1960), and Jacobson, Trumbo, Cheek and Nangle (1959).

[22] The 14 jobs aspects are: amount of work variety: amount of work; degree of work accuracy; control over pace of work; importance of work for firms; amount of supervision received; amount of skill work requires; amount of responsibility; amount of planning; amount of judgement; interest of work; job security; promotion chances; and pay.

been changed by the computer; in firm B, although some computer impact was reported by 79% of staff in affected departments, substantial changes were confined to a minority of employees. Similarly, in Carters, 32% of the clerks said their jobs had not been changed at all, and 76% said that less than four aspects of their jobs had been changed.

In contrast, in Royal Exchange, only 10% said their jobs had not been affected at all, more than a half reported a change in four or more aspects of their jobs, and 20% reported a considerable change in their jobs (more than 7 job aspects had changed). Three aspects of work were said to have changed by more than 50% of clerks in one or more of the firms. These were: interest of work (Royal Exchange), amount of work (Royal Exchange, Carters, American Firm A) and variety of work (American Firm A, American Firm B). We do not distinguish here between clerks who said the factor had increased and those who said it had decreased.

TABLE 27 *Aspects of work which more than half the clerks in a particular firm said had changed*

Royal Exchange	CARTERS	American Firm A	American Firm B
Interest of work (59%)	—	—	—
Amount of work (53%)	(62%)	(67%)	—
—	—	—	Variety (57%)

The table below shows the job aspects which clerks in the two British firms and in American Firm A believed to have changed most and the direction the change took. Clerks are treated as a homogeneous group and there is no separation into departments.

Greater accuracy and more work were demanded by the new system in Royal Exchange and American Firm A. In Carters, work and planning increased, but only for a sixth of the clerks. Work decreased for nearly a half of Carters clerks and nearly a third believed that security of employment had been adversely affected by the computer. Interest and variety decreased for many of the Royal Exchange clerks. Few people in American Firm A found that anything decreased very much, although nineteen per cent found they now had less work to do.

There were considerable differences in the overall direction the

TABLE 28 *Clerks' perceptions of the ways in which their jobs had changed most*[23]

	Royal Exchange	CARTERS	American Firm A
	Accuracy (42%)	—	Accuracy (59%)
	Work (38%)	Work (19%)	Work (48%)
	Planning (36%)	Planning (19%)	—
	—	—	Variety (47%)
MORE	—	—	Responsibility (47%)
	—	Control over pace (15%)	—
	Interest (38%)	—	—
	—	Work (43%)	Work (19%)
	—	Security (29%)	Security (15%)
LESS	Variety (29%)	—	Variety (14%)
	Control over pace (21%)	Control over pace (24%)	—
	—	—	Promotion (14%)

changes in job content took in each firm. More factors increased than decreased in American Firm A and the American research team concluded that this should lead to an increase in work satisfaction—implying that in the 14 job aspects they examined 'more means better'. In Royal Exchange there was a tendency for job aspects to increase rather than decrease, but the difference was small and the American hypothesis was not supported, for work satisfaction certainly did not increase. In Carters, job aspects decreased more than they increased, but work satisfaction remained the same.

Our results, then, do not support the American suggestion that 'more' of the 14 job factors generally means more job satisfaction and 'less' means less satisfaction. The important point seems to be the area of work where an increase or decrease occurs. In Royal

[23] Employees of American firm C whose jobs had changed reported more variety 26%, more work 21%, and more interest 19%. American firm B employees were only asked whether or not their job had changed, but not the direction of the change.

Exchange it can be suggested that job satisfaction would not increase as a result of the changes in job content because the two aspects of work that intelligent girls would consider most important (interest and variety) were the ones that decreased most, whereas accuracy and amount of work, neither likely to provide much satisfaction, increased. A considerable amount of research has taken place on monotony in work and, although the evidence is not conclusive, interest and variety appear to be important elements in job satisfaction for many people, particularly the more intelligent and the young. It therefore seems appropriate to stress again the point we made earlier, that the bank was recruiting girls of too high an intellectual calibre for the new simple machine jobs. In Carters, where there were lower educational standards, monotony was not a problem. It is possible, of course, that personality factors are more important than intelligence in the ability to tolerate monotony. However, it seems unlikely that Carters were recruiting girls of one personality type while Royal Exchange selected those of another.

The increased need for accuracy mentioned most frequently by Royal Exchange and American Firm A clerks would appear to be a source of stress rather than satisfaction. Mann and Williams (1962) refer to the increased risk associated with automation in the office. Errors cost time and money. Punching errors, for example, if not identified before the tape left the branch, caused the computer to stop and meant that the data could not be processed until the following day. It was also easier to detect the girl who had made the error so that there was less chance of escaping blame.

Let us now establish whether ways in which work changed for clerks in the computer areas differ from those for clerks in departments which had work transferred to the computer. In Carters we will class girls who worked in the Hollerith room and in the computer centre as 'computer area' personnel. Clerks from elsewhere in the Accounts Department will go into the 'affected department' category.[24] In American Firm A the research team also distinguished between computer area clerks and clerks in affected departments. In Royal Exchange we were unable to interview girls in the computer centre and so all our clerks go into the 'affected department' category.

[24] There were only seven of these who had experienced both pre- and post-change situations.

TABLE 29 *Job aspects which computer area clerks said had changed most*

	Royal Exchange	CARTERS		American Firm A	
Computer Area		Planning	(28%)	—	
		—		Variety	(74%)
		Work	(21%)	Work	(74%)
MORE	No data	—		Skill	(74%)
		Judgement	(21%)	—	
		Importance of work	(21%)	—	
		Work	(50%)	—	
		Security	(36%)	—	
LESS		Variety	(28%)	—	
		Control over pace of work	(28%)	Control over pace of work	(6%)

Variety, amount of work and skill went up considerably in American Firm A's computer area, but this was not true of Carters. Here the computer's most noticeable effect on jobs was less work, less perceived security and less variety and control over the pace of work. The effects of automation on the job content of computer area staff were, therefore, more desirable in the American firm than in the British.

The following table shows how clerks in 'affected' departments in Royal Exchange and American Firm A thought the computer had changed their jobs. Carters is excluded from this table because few of the small group of clerks working outside the computer area thought their jobs had altered in any way.

Similar work changes were experienced by clerks working in the affected departments of the British and American firms. The only important difference is the reduction in the interest of work, found by 38% of the British bank clerks.

The fact that we were considering only computer area staff in Carters and only clerks in affected departments in Royal Exchange makes it difficult to generalise on the changes in job content in these two different kinds of situation. But it does seem that the more desirable aspects of work for most people—variety, judgement, skill, etc.—tended to be increased for computer personnel and reduced for staff losing work to the computer. Certainly the

TABLE 30 *Job aspects which clerks in affected departments said had changed most*

Affected depts.	Royal Exchange		American Firm A	
MORE	Accuracy	(44%)	Accuracy	(56%)
	Work	(38%)	Work	(39%)
	—		Responsibility	(39%)
	Planning	(36%)	—	
LESS	Interest	(38%)	—	
	—		Work	(24%)
	Variety	(29%)	Variety	(20%)
	Control over pace of work	(21%)	—	
	—		Security	(20%)

American team believed this to be the case (Faunce, 1962). They found that their clerks liked the changes arising out of the new system more than they disliked them, but, whereas responses indicating approval outnumbered those indicating disapproval in the computer area, clerks in affected departments showed a mixed pattern of response. The American team concluded that employees in a computer area, which gains work, will like change more than employees in affected departments which lose work and have to adjust to new methods. Our data would support this conclusion.

If we now turn our attention from the aspects of work which changed to the aspects of work which did not change, we can include data from American Firm B where employees were given eleven of the fourteen job aspects and asked 'Did the change-over to the computer affect this aspect of your job?'[25]

A majority of the clerks in all the firms were in agreement in three of the four work aspects they said had changed least. These were: pay, the amount of supervision, and promotion chances. There was some difference of opinion over the fourth item. Royal Exchange branches and American Firm A clerks said security of employment; Carters clerks said skill, and American Firm B clerks said the ability to control the pace of one's work.

[25] The questions on importance of work for firm, amount of planning work requires, and amount of judgement, were omitted.

As a result of their analysis, the American research team came to a number of conclusions (Hardin, 1960a.). First, that computers cause some changes relevant to employee satisfaction, but that these changes do not constitute an acceleration or a reversal of general trends. This form of automation, revolutionary though it may be technologically, causes changes in work environment and job satisfaction very similar to those which occur normally and without computer automation.

Our data for Royal Exchange do not support this conclusion, but, on the contrary, indicate that a change-over from a largely manual and personal work system to automation did lead to a decrease in work interest.[26] We have no evidence that the interest of work was declining generally throughout the bank due to normal day-to-day changes in the office routine; our North office data suggest that change in non-affected departments was minimal. The formalisation of banking procedures would in fact make this kind of change very gradual, if it occurred at all. The change at Carters, where there was an existing punched card installation, had little impact on the computer area apart from an increased threat of redundancy; outside the computer area insufficient staff were affected for any conclusions to be drawn.

The second conclusion of the American research team was that 'the process of installing a computer and of adapting work methods to it may cause noticeable, though presumably temporary, dissatisfaction unless management handles the conversion with great skill' (Hardin, 1960). The data we presented in earlier chapters would support this hypothesis although we have no evidence that dissatisfaction will be only temporary. On the contrary, it can be suggested that if staff believe they have suffered a reduction in job interest and job security, unless compensations are provided to make up for these defects, there is no reason for them to change their attitudes from dissatisfaction to satisfaction.

If we attempt to summarise this last section, it becomes clear that it is impossible to identify any effects that were common to all the clerks in all the firms. This suggests that the kind of impact a computer has on work is conditioned to a large extent by the situation into which the machine is introduced. This situation embraces not only the nature of work performed before and after

[26] We would agree that a centralised punched card system might have had the same effect.

E.D.P. but also the attitudes of clerks to their old and new jobs—and many of these attitudes seem very subjective. One cannot therefore, say to a manager 'if your firm gets a computer then jobs in your offices will be changed in the following way'. For example, in Royal Exchange and Carters we found widely differing views on the status of the same jobs, and in the British and American firms there was little consensus of opinion on how E.D.P. had altered the different elements making up the clerks' jobs. Only two generalisations seem possible. The first is that work content will be more significantly altered if the change to E.D.P. is from a manual rather than a punched card system of work. The second is that staff in the computer area who see themselves as the agents of change will view changes in their jobs more favourably than clerks in affected departments who are on the receiving end.

CHAPTER 10

The management of change

The literature of industrial sociology and social psychology contains many references to 'resistance to change' and the implication behind some of these writings appears to be that this is a natural human condition: we all have a tendency to wish to preserve the status quo and to dislike innovation. This would seem to be an over-simplification. Whether we resist, welcome or are neutral towards change will depend on the interaction of many different and complex variables, some of which lie within ourselves in our personalities and characteristics; others are part of the environment in which we work and live.

This means that any major change in technology within an organisation will evoke different reactions in the individuals affected, depending on the nature of the variables influencing the behaviour and attitudes of those individuals. Because of this complex interaction process, the social consequences of a technological innovation such as a computer can only be accurately predicted up to a point, on the basis of other people's experience. Although staff problems observed in other firms can act as a guide, each firm must identify the particular factors motivating the behaviour of its own staff and decide which variables in the change situation are likely to produce positive or negative reactions.

Variables conditioning attitudes will lie both inside and outside the firm. For example, variables operating within the work situation may be the kinds of satisfaction an employee wishes to obtain from his job and his degree of identification with the firm. Outside variables could be the amount and content of the propaganda about technology put out by press and television at a particular moment of time, the nature of a person's family responsibilities, and the comments of his friends and relatives.

The variables most likely to have a considerable influence on staff reactions have emerged from the preceding analysis of our research data. These are, firstly, the consequences of the change—what actually happens to the work situation and to the tasks of individuals and groups involved in it. These consequences can be divided into *actual* and *perceived*—what really does happen and what people think has happened. Sometimes these two things are the same, but sometimes they are not. In the last chapter we discussed the actual and perceived consequences of the new system for clerks in Royal Exchange and Carters, and in Chapters 5 and 6 we showed how the inaccurate perceptions of North Office clerks produced a distorted picture of the changes that had taken place in the branches. Secondly, a major variable will be the staff themselves, their characteristics, their pre-conceived ideas, their goals and aspirations and how the change is seen as helping or hindering the achievement of these. Thirdly, the kind of situation in which the change takes place will be important—the existing climate of relationships between staff and supervision, and within and between occupational and work groups. Fourthly, there is the way the change is handled by management: the kinds of strategy and technique that are used for introducing the change and inducing staff to accept it. In this chapter we shall consider in turn the last three variables; the clerk himself, the situation in which the change takes place and the management of change.

The reactions of clerks to change

Let us start by reminding ourselves of the reactions of the clerks in our two case study firms and comparing these with the attitudes of similarly affected clerks in other situations. Relatively few of the Royal Exchange or Carters clerks had any great anxiety about automation before the change, but the fears of those who did were quite different in each firm. In Royal Exchange the suggestion most frequently made by women was that the computer would lead to a decline in the interest of work (20%); in Carters it was that automation would lead to redundancy (22%). Men, in contrast, tended to give neutral or approving answers to the question on how computers would affect clerical work. The three Carters male clerks all said 'I don't know'; the male bank clerks described likely changes in work organisation (28%) or said work would become easier or more interesting (22%). Carters clerks were generally much vaguer than those at the bank. 60% said either that automation would have no effect or that they did not know what effect it would have, compared with 20% in the bank. It seemed that, whereas the bank clerks had thought about automation and indeed had some clear and accurate ideas as to its possible consequences, the clerks at Carters were still, at the time of the pre-change interviews, largely in ignorance of what a computer would involve. This ignorance was, of course, in part due to the efforts of Carters' management to keep the computer a secret. Information was more readily available in the bank because, with the exception of King George's Place branch, all the interviews were carried out after the computer had arrived and when rumours were spreading from King George's Place to the other branches.

After the change we found, at Carters, that although there was still anxiety about redundancy, the girls approved of the new system, and jobs on machines were even more sought after than before the computer arrived. But in Royal Exchange girl machinists were very unhappy with their new jobs; they complained that these had lost interest and now resembled factory work. Also, although male clerks had been little affected by the computer, there was some evidence of a drop in their morale and a belief that promotion prospects had deteriorated. In North Office, the last branch to be investigated, there were gloomy impressions of the kind of changes that had occurred in the affected branches and signs that bank staff had at last begun to associate computers with redundancy. In both the E.D.P. branches and North Office, the majority of clerks believed their new computer installation had brought advantages to the bank, although not to the individual clerk or to the customer.

The quite different types of reaction that we found in Royal Exchange and Carters have also been observed by other research workers studying office automation. Mann (1962) comments on the lack of anxiety amongst clerks in the American insurance firms studied by the team from Michigan State University. He compares this with his own study of a light and power company, where clerks were very worried about the possibility of redundancy despite management assurances that this would not occur. He relates these differing reactions to the extent of the office automation and concludes 'how sensitive employees in the office are to these problems of potential technological change appears to vary directly with the size of computer conversion that their organisation has experienced'. Our research findings would lead us to qualify this statement and to suggest that, although the extent of E.D.P. must be an important variable, as it will control the amount of change that occurs, other variables present in the change situation may be equally important in conditioning attitudes.

The accepting attitudes of the clerks in Carters and in the Michigan insurance companies did not seem to be a typical response to office automation. Seven British firms were visited during the course of our research in order to see how representative our case study data were of other firms' experience. Most of these firms told us that staff suffered a great deal of anxiety when first told that a computer had been purchased and that this anxiety occurred whether communication had been adequate or not. The only firms expressing the view that anxiety and unrest were not serious problems were those which had ordered but not yet received their computers. One retail organisation found that 'expressed' fears were greatest over a short period when running in parallel stopped. This company also encountered a reluctance to co-operate on the part of affected departments and, in fact, the computer centre staff had themselves to get out all the input records for the computer, as the departments holding these were unwilling to do the necessary work. Some firms had staff who were convinced that the computer was not capable of producing accurate data and, in the early days of computer take-over, feelings of insecurity often made them try to demonstrate that this was in fact the case.

This sort of anxiety was found equally in many American companies. Mann (1962) says 'employees now keep their company's policy statement assuring no loss of employment because of technological change carefully filed where they can get it at a moment's notice'. Siegman and Karsh (1962) describe an American company where, as in our British retail organisation, departments refused to supply programmers with basic information. Mann and Williams (1960) found the computer being used as a convenient scapegoat, 'more and more mistakes were attributed to the machine . . . it was an unknown and unpredictable monster'.

The drop in morale that we observed with the bank clerks was also found elsewhere. Mann and Williams (1962) studied one firm where 'the proportion who felt their future looked much worse or somewhat worse increased significantly from before to after the change-over period'. Kushner (1963) claims that 'the period of change is a trying one for employee morale. Most of the time the individual employee must operate on faith—faith that demands made upon him will be rewarded in the future'.

Generally speaking then, it seems that in most firms the introduction of a computer can be expected to cause alarm and anxiety in clerks and management, especially during and immediately after the change-over. Situations where anxiety may be less of a problem are found in those firms which, like Carters, employ mainly very young women clerks whose job investment is low; where the impact of the new system is reduced because of the existence of a punched card installation, and where the bulk of the procedures to be transferred to the computer are already processed mechanically. Even in Carters, however, although the girl clerks were favourably disposed towards the new system, they were still worried about the possibility of future redundancy.

The individual who experiences the change

Let us now consider the extent to which the personal characteristics, motivation, goals and job attitudes of the individual clerk may affect the degree to which he accepts or rejects change.

VARIABLES AFFECTING ATTITUDES TO CHANGE (1)

Variables lying within the individual

Social characteristics	*Personality*	*Goals*	*Job attitudes*
Sex	Introvert/ Extrovert	Predominantly work centred.	Degree of satisfaction with pre-change job.
Age	Rigid/ Adaptable	Predominantly outside the work situation	Degree of job investment (length of service etc.)
Social Class	Dependent/ Independent		
Education	Passive/ Initiating		

SEX

Our evidence suggests that men are more likely to be anxious about technical change than women. However, the reason for this is not that sex differences cause different reactions to change stimuli, but that most men do not have the same goals and aspirations as most women. We will therefore discuss the attitudes of men and women in the section on 'goals'.

AGE

It is commonly thought that age is an important factor—the older people get the less likely they are to welcome new ideas or methods. In Royal Exchange we did find that young male clerks were more likely to favour change than older clerks but the dividing line distinguishing young from old was around twenty-one. This suggested that it was not so much age in itself that affected attitudes as degree of job investment. The twenty to thirty-year-old clerk was under way with his career and also acquiring family responsibilities. Any change in a hitherto predictable situation was threatening to him, for he was not close enough to promotion to be certain that he would achieve this.

We interviewed very few male clerks over thirty in the branches, but in North Office there were a number in the thirty to forty year age range and, like the young clerks, they tended to the pro- rather than anti-change. Again, this group were at a less critical stage of their careers; either they had given up any idea of promotion or they knew that promotion was reasonably certain within a year or two whether office automation took place or not. American research supports our conclusion that age itself is not a very important factor conditioning attitudes to change. Miller (1960), after studying the introduction of computers into six companies, found that age had no adverse effect upon the employee's attitude towards the methods change. In three of his six companies, the older people were more favourably disposed towards it than the younger ones. Perhaps a note of caution is appropriate here, however. All our clerks were under forty, whereas 'psychological ageing'—bodily deterioration and its effect on the mind—does not make its full impact until middle age.

SOCIAL CLASS

The Michigan State University research team found a relationship between social class and attitudes to change; a higher social class background (using father's occupation as the measure of this) being associated with more favourable attitudes to change (Faunce, 1960). It will be remembered that we also tested this variable in Royal Exchange and Carters but were unable to find any supporting evidence for the American conclusion. If anything, we suspected that in our two firms, social class status was having an influence in the opposite direction. The Royal Exchange girl clerks, who came principally from middle-class homes, reacted against the new machine punching jobs because they associated them with factory work. This similarity between punching and factory work did not trouble the Carters girls with their working-class backgrounds. Many had parents who worked on the shop floor and some had considered this kind of employment for themselves.

EDUCATION

The American study also suggested that a higher education might go with pro-change attitudes (Faunce, 1960). We could find no relationship between grammar and secondary modern school educations and attitudes to change, although in educational attainment we again found a trend that contradicted the American hypothesis. In the pre-change studies, Royal Exchange girls with G.C.E. were more anti-change than other girls. These girls had entered banking believing it to be an interesting job, worthy of their academic qualifications. They became afraid, with some justification, that the computer would depress the nature of their work and make their knowledge and qualifications irrelevant.

PERSONALITY

Besides social characteristics such as sex, age, class and education, each individual has a personality make-up which has been developing since childhood. This, too, is likely to affect attitudes to change. There is some evidence that people who are security minded and rigid in outlook are less adaptable than those who rate security less highly and who show flexibility in their attitudes.

As we did not give our clerks personality tests we have no direct knowledge of their personality characteristics. But the nature of banking with its security, routine and life-long commitment would seem likely to attract the more passive and less adventurous type of individual.[1] McMurray (1958) referring to American bank clerks, suggests that the 'right type' of bank employee is 'neat, clean, well groomed, comes from a good family, speaks quietly, uses good grammar and is properly deferential to authority figures. . . . He is the submissive type who has always done what he is told. . . . Men and women who conform to this stereotype have been sought for and hired by banks for many years.' The fact that a third of the Royal Exchange clerks said they had been influenced to take up banking by a relative or friend, and that this was particularly true of grammar school girls who would have a wide choice of career, might indicate a dependent nature and a seeking after security although we have no other direct evidence that would support this point.

In our earlier analysis we found that the clerk who was not favourably disposed towards the new system of work was well adjusted to banking; he liked his job and believed that his promotion prospects were good. In contrast, the clerk who did not like his job very much and who was not confident of promotion, was in favour of the computer. This suggests that the kind of personality who is attracted to banking and who, in the past at least, has made a success of it, may find the new developments especially hard to accept. Conformity and submission to authority, which banking has traditionally demanded of its employees, are quite acceptable to him, but adjustment to change is less so. We hypothesise that the bank clerk (and security minded individuals in other occupations) will accept change without making trouble because it is in his nature to do what his superiors tell him. Nevertheless he is likely to find it more worrying and unpleasant than, for example, the more extroverted advertising man or sales executive.

NEEDS AND GOALS

Another important influence on attitudes to change will be the motivation and personal goals of the individuals affected by the change. What are they striving for in life and will they see the new

[1] Some British motor insurance companies apparently believe this, as there has recently been a suggestion that bank clerks shall be offered particularly favourable policy terms.

situation as likely to help or hinder their achievement of these objectives? March and Simon (1958) suggest that we try to satisfy our needs through identification with different aspects of our environment and that there are four principal targets for this identification. First, organisations external to the focal organisation, such as our families and our social and professional groups. Second, the focal organisation itself—for our clerks this would be Royal Exchange or Carters. Third, the work activities involved in the job; and fourth, the sub-groups within the focal organisation, such as social and task groups in the work situation.

Because we do not all achieve satisfaction in similar ways, individuals will vary in their degree of identification with these targets. Some people will have their principal goals centred on their families or their social life and work will become less an end in itself and more a way of securing the economic means to achieve domestic or social goals. For others there will be a strong tendency to satisfy personal needs and ambitions through the medium of the job, and March and Simon (1958) suggest that when job characteristics permit such satisfaction people will be very identified with their work. We can take this hypothesis a stage further and suggest that attitudes towards change are positively related to the importance of work as a medium for achieving desired goals.

This means that the person whose major goals lie outside the work situation will be less concerned about what takes place there, and should therefore be less concerned about change than the person who uses his job as the principal mechanism for achieving his ambitions. Most of us, of course, do not fit neatly into either of these categories. Some of our goals lie outside the work situation, some within it. If, for any reason, we have chosen the wrong career and find we cannot satisfy our needs through work, then we shall either become very frustrated or else we shall replace work goals with social goals and lose interest in our jobs. Similarly, if our family or social lives are unsatisfactory, work will become of compensatory importance and more of our goals will be centred on it. Needs and goals are not, of course, constant, but will be continually changing through life; as each goal is attained it will be replaced by another.

If we accept that the nature and focus of our goals influence our attitudes, we would expect men to be more anxious about change than women. This follows from the fact that men, as a rule, have a greater degree of job investment than women, and therefore more work centred goals. We must, of course, exclude career women from this generalisation.

Broadly, our research findings support this conclusion. An important goal for the majority of Royal Exchange male clerks was to do well in their jobs and achieve promotion (58% of the male clerks had plans of this kind for their future careers). But, whereas before the change these clerks generally approved of the idea of a computer, afterwards, when the new system's implications became clearer, significantly fewer said that the computer's effects had been favourable. Also, branch men showed signs of a decrease in job satisfaction after the change and were less confident about their promotion chances.

Women clerks in Royal Exchange were far less work orientated than men; for many the principal goal in work was pleasant social relationships. Only 12% had any plans for their future careers and all of these plans involved leaving the bank, either for marriage or another kind of job. Labour turnover amongst this group was high and, unlike the men, they were not tied to banking but could look elsewhere for new jobs. In view of the girls' lack of involvement with work, and the easy escape route into other employment, one would expect that the bank's automation plans would disturb the girls less than the men and this proved to be the case. The girl machinists were, of course, an exception here: whereas the job content of the male clerks and of those female clerks on non-machining jobs had changed little, the machinists had suffered increased monotony and routinisation of work. They had therefore encountered the objective rather than the subjective consequences of automation and this had influenced their attitudes. It seems probable that many of these girls will leave and that the bank will find this job difficult to fill in the future.

Carters' girl clerks appeared rather more interested in work than Royal Exchange girl clerks and it was the Hollerith group who showed most job satisfaction. Insofar as the goals of this group lay within the work situation, the new computer installation would further these by giving the girls additional status through their membership of the innovating group. The computer's only threat was the possibility of redundancy in the future, and the Hollerith girls, who suggested this, were fairly confident that their own jobs were secure.

Within the clerical ranks then, the computer presented the greatest potential threat to the Royal Exchange male clerks who had many work centred goals, a high degree of job investment and, because bank clerks cannot transfer from one bank to another, very restricted job mobility.

We have little systematic information on the goals of affected managers in the two organisations, although informal interviews suggested that these were status, promotion and, in Royal Exchange, the prestige resulting from a smoothly functioning branch and good customer relations. Managers could be expected to be even more work centred than their clerks, and the Royal Exchange managers expressed anxiety about the effects of the new system on their access to accounts information and on customer relationships. For this group too, the computer was a threat, and likely to hinder the achievement of some of their personal objectives.

The two groups of programmers were in a very different situation. Even more than Carters' Hollerith girls, they saw the new system as a means for achieving increased work satisfaction and, because of this, they had a vested interest in the continuance of change and innovation. They told us that they liked the challenging and varied nature of the work, and the fact that it was so new. It also offered them career opportunities in a type of work where skilled technicians would be in short supply for some considerable time to come.

Our evidence suggests that the goals of the different occupational groups in Royal Exchange and Carters did affect their attitudes to change. Royal Exchange male clerks and managers, who had important goals located in the work situation, saw the computer and the new system of work as threatening to these, and were therefore disturbed and apprehensive about the change. Royal Exchange and Carters girl clerks, who were not seeking much more than a pleasant social situation and work that was not too dull, did not see the computer as greatly affecting their goals and were less anxious; although they were not completely unperturbed as their references to redundancy and the likelihood of a reduction in job interest indicated. Royal Exchange and Carters programmers, in contrast, saw the computers as assisting the attainment of their goals. They could use automation to achieve increased job satisfaction and better career prospects.

Other research workers support our hypothesis that an individual's needs and goals are an important factor influencing his attitudes to change. Trumbo (1961) for example, states that 'an employee's response to change is probably conditioned by his perception of the way in which the effects of change relate to his needs. If change is to be accepted—its effects must be seen as generally more rewarding than unrewarding.' McGregor (1960) and Mann and Neff (1961), emphasise the necessity for implementing change in such a way that individuals, in achieving their own goals, also help the organisation to achieve its goals.

JOB ATTITUDES

A last factor that needs to be considered in this section is the influence on attitudes to change of an individual's degree of satisfaction with his existing job. Here again we discerned a relationship, for clerks who found their present jobs interesting did not look favourably on change, while clerks who did not like anything much about their jobs welcomed it. It can be suggested that employees who do not get much satisfaction from their jobs may see change as bringing an improvement. This leads to the most interesting hypothesis that change may be more readily accepted in situations were there is little pre-change job satisfaction. Organisations with employees who enjoy their jobs may find that the introduction of a technical innovation, such as a computer, causes considerable stress and anxiety. This suggestion could usefully be investigated further as it contradicts some of the points we shall be making in the next section.

The situation in which the change takes place

VARIABLES AFFECTING ATTITUDES TO CHANGE (2)

Variables in the situation in which the change takes place

Past Policy	*Propaganda*	*Relationships*
The way the firm is known to have handled change in the past.	The amount and kind of information circulating about the proposed change or similar changes in other firms	Existing: Between managers and departments. Between employees and the firm. Between supervisors and subordinates. Between colleagues

Another set of variables likely to affect attitudes to change are located in the situation in which the change takes place. One of these is the way the firm has handled change in the past; another, the sort of propaganda that is in circulation relevant to the kind of change the firm is proposing to introduce. Both of these will give employees pre-conceived ideas and expectations about what is going to happen. A third, and most important influence, is the existing climate of relationships within the firm. This is a product of the relationship between one department and another; of the employee's degree of identification with the organisation and its goals, and of his relationship with supervision and his colleagues.

PAST POLICY ON CHANGE

In Royal Exchange and Carters, neither group of clerks had any pre-conceived ideas derived from methods used for handling change in the past. Both firms were very slow moving and had had no major changes in office procedure within the employment span of most of the clerks in our sample. Royal Exchange had had its accounting machines for some time and Carters' Hollerith installation was an old one. None of the clerks we interviewed in the Hollerith section had been there when it was introduced. This meant that the clerks did not enter the change period expecting either that the firm would handle things well 'because it always does' or 'things will be bungled as usual'. They were quite neutral in their expectations.

PROPAGANDA

The principal agent moulding and creating ideas about the change was therefore the verbal exchange of opinion and it will be remembered that, with the exception of the first branches to go on the computer in Royal Exchange, rumour was the most common method for aquiring information.

In Carters, although there was very little pre-knowledge about automation, the absence of this did not appear to create advance prejudice against the computer. In Royal Exchange, in contrast, rumours of technical difficulties spread from branch to branch and did cause adverse pre-change attitudes to develop in both clerks and managers. Our study of North Office showed how staff in a branch which was never transferred to the computer had pre-conceived ideas of a very gloomy kind. This whole question of communication will be discussed in detail further on when we consider the management of change.

RELATIONSHIPS

In both firms staff relationships were good and there was no evidence of any major pre-change discontent. Male clerks, particularly those over 21 years of age, seemed identified with the bank and banking although we have shown that this identification was more likely to lead to unfavourable than to favourable attitudes to change. The fact that male clerks did identify in this way meant that although they were anxious about the consequences of change, they were quite prepared to accept it and believed that it was in the bank's interest to adopt the new automated system.

In neither firm did the female clerks show much evidence of organisational identification, with the exception of a few career women in the bank. This could have been an advantage to management as it meant that women were relatively unconcerned about the changes that were taking place.

Relationships between departments seemed harmonious in Carters and were virtually non-existent in Royal Exchange as the individual branches were to a large extent autonomous. Departmental problems occurred only during the change-over when both firms had difficulty in maintaining good relations between affected units and the computer centres.

The pre-change relationships existing between management and employees, particularly immediate supervision and their subordinates, must be an important influence on the way change is received. During a change period the ambiguity of the situation makes employees insecure and uncertain about the future. They are therefore especially dependent on their supervisors for information and re-assurance. If the supervisor is not accustomed to relating in this way with his subordinates, then he will not be able to give them the help and guidance they are looking for and employee morale is likely to drop. More is required of the supervisor than re-assurance and guidance, however. One of the most important aspects of his role before the change gets under way is that of motivating and persuading his subordinates to look kindly on the proposed innovation. Barnard (1952) lists the five essential qualities of leadership as (1) vitality and endurance; (2) decisiveness; (3) persuasiveness; (4) responsibility, and (5) intellectual capacity. We would suggest that quality (3), that of persuasiveness, becomes extremely important when a firm is about to introduce a technical change.

Mann and Hoffman (1960) believe that all supervisors must have administrative competence, technical competence and human relations competence; the latter being used to integrate organisational objectives with the needs of the individual. While the need for technical skills may grow as the change develops, human relations skills are undoubtedly of the greatest importance before the change is implemented. It is essential to set the change against a background of trust and goodwill between management and employees. This mutual trust and goodwill certainly existed in Royal Exchange and Carters but the human relations skills of the branch and departmental managers were undoubtedly inhibited by their own lack of information about what was going to take place, and by their own personal misgivings. It is difficult for supervisors to support changes if they are not themselves entirely convinced of the wisdom of these.

Existing relationships within the work force must also play an important part in the acceptance or change. Technical change normally causes a redivision of labour and a breakdown of existing work group organisation. This means that employees working in cohesive and well integrated groups and, like our Royal Exchange girls, placing a great deal of importance on pleasant social relationships, may see change as a threat to the social aspects of their jobs. However, in neither Royal Exchange nor Carters was there much evidence that clerks were worried about a disturbance in existing group relationships. Four of the five Royal Exchange branches, and Carters Accounts Department, were quite small and this meant that even though work was re-organised, girls would still be able to work near their friends. The exception here was Royal Exchange North Office, where we interviewed 54 clerks and where over half of the female clerks in the sample said that social relationships were the job aspect they liked most. This was the only branch in the pre-change study where women clerks predicted the displacement of staff and suggested that the computer would cause girls to be transferred to other branches. We conclude that women staff are more likely to be apprehensive about the effects of change on social relationships than men, and that apprehension will be greater in a large office where the impact of a new work system is likely to break up an existing group structure.

The way the change is managed

VARIABLES AFFECTING ATTITUDES TO CHANGE (3)

Variables in the policies adopted by management for introducing the change

Communication (formal & informal)	*Consultation & participation*	*Timing*	*Rewards*
To what extent is there: Motivation to communicate. An efficient communication system. Appropriate communication techniques.	Will this be effective? If yes, who shall be involved in it and to what extent?	Is time allowed for adjustment to the idea of change, and for the prevention of stress.	Some physical and mental stress is inevitable. Is it being adequately rewarded?

Yet a third set of variables influencing attitudes to change are the policies and strategies adopted by management to get the innovation working and accepted. These are the variables over which a firm has the greatest short run control. It can do little to alter those variables lying within the individual, and the kind of social situation in which the change takes place is something which usually has developed slowly over many years. Before we discuss the pros and cons of different approaches to the management of change, one most important point must be stressed again. This is that it is impossible to produce a blue-print for successful change. An approach that is appropriate in one firm may be quite inappropriate in another. Policies must be developed in terms of the needs of particular situations. These needs can only be ascertained through painstaking analysis of the two sets of variables we have already discussed, the characteristics of the people working in the firm, and their motivation and goals; and the nature of the social environment into which the change is to be introduced. This kind of analysis is often done instinctively and unconsciously by management. Sociologists suggest that it can be done intellectually and systematically. Lupton (1965) believes it would pay firms embarking on major technical change to bring sociologists in to do it for them.

Let us now examine some of the mechanisms for aiding acceptance which need to be considered when devising a strategy for change. The principal ones are communication, consultation and participation, timing, rewards and compensation.

COMMUNICATION

What is communication? In the older management text books it is presented as a set of formal techniques operating in a one way and downwards direction. Managers are encouraged to become better writers or better verbalisers in order ot get their ideas across to subordinates who, apparently, only need to learn in order to understand and accept. Today we know better than this. Communication is increasingly recognised to be a continuous interaction process which, as Cook (1951) has pointed out, usually reflects the nature of inter-personal relationships within an organisation. It is less a moulder of attitudes than a product of these. If communication in a firm is unsatisfactory, this may be not so much because managers are poor writers and talkers as because people do not get on well with each other and therefore do not want to pass on information. Even if relationships are good, knowledge often means power and both organisations and individuals are tempted to restrict the free flow of information to maintain or extend their own power position. Weber suggests that bureaucratic organisations lean towards secrecy; that bureaucracy welcomes a poorly informed and hence a powerless proletariat, at least in so far as ignorance agrees with the bureaucracy's interests (Gerth and Mills, 1948). We find also that managers are often reluctant to keep workers and shop stewards informed of changes that are being planned, because this knowledge would enable shop floor groups to take counter action. Similarly, in our computer researches we found that a number of firms, including Carters, decided on secrecy as a deliberate policy and tried to present their employees with a *fait accompli*. Management suggested that this was to avoid causing premature anxiety in staff who were likely to be affected. This may have been true, but in some firms it seemed more likely that the real reason for secrecy was the desire to prevent staff taking evasive action—by looking for other jobs, for example.

Communication takes many forms, and we have seen in Carters and Royal Exchange how informal communication, notably rumour, was by far the most important medium for disseminating information. Cook expresses surprise that so little attention has been paid to this method of communication, in view of its general use in society. Even if formal communication is very good, the bulk of information passing will take place informally. The reason for the emphasis on formal communication in industry is, of course, that management believes it can exercise control over this, whereas informal communication is less easy to manipulate.

In discussions of communication great stress is always placed on verbal and written aspects and it is not generally recognised that a great deal of communicating is done through behaviour. In Carters, management tried to keep news of the computer from staff although it was not very successful in doing so. But, by their very silence, they may have communicated certain ideas to staff. For example, that 'we have something to hide', 'the new developments will make you redundant and so we are not telling you about them', etc., As it happened, Carters management was fortunate and its silence was not interpreted in this way by the girl clerks. This leads us to another point about communication; that the way a message is interpreted depends on the variables we have discussed earlier in this chapter—the individual's social characteristics, motivation and goals, and the nature of the relationships present in the situation in which the communication takes place. Communication about change is especially prone to widely differing interpretations, depending on the particular interests and anxieties of the groups receiving the messages.

Effective communication, that is, communication that reaches its destination undistorted, and is interpreted in reasonably similar ways by groups and individuals on the receiving end, has a number of organisational advantages. First, it is a good conflict resolving mechanism, it helps to eliminate misunderstandings by keeping people in the picture and letting them know what is going on. Second, it leads to improved problem solving. Because most of us approach problems in different ways, communication enables ideas to be evaluated and difficulties identified (Blau and Scott, 1963). Third, it can lead to increased acceptance of new ideas. If employees are throughly acquainted with the change objectives and procedures they are more likely to accept change—always providing that it does not conflict with their personal needs and goals.

However, effective communication is not the panacea for industrial troubles that it is sometimes made out to be. On the contrary it can sometimes increase a firm's problems. If there is conflict between groups and individuals then communication may increase this by making the areas of difference more explicit. Similarly, if a group feels threatened by a proposed change, telling the members about the change will not make them readier to accept it, but may increase resistance. This is not to condone a policy of secrecy, for, ethically, staff have a right to know what is going to happen to them. But it does mean that in these circumstances mechanisms other than communication will have to be adopted to secure the acceptance of change.

Another disadvantage of effective communication is that, while improving problem solving, it also impedes co-ordination. Unrestricted communication creates a battleground of ideas and may make it difficult to reach agreement (Blau and Scott, 1963). This is particularly true when the individuals communicating are of equal status.

Communication needs in times of change are rather different from those found in periods of stability. There is an increased demand for information because staff will be worried by the ambiguity and uncertainty of a changing situation. There will also be a demand for rapid and accurate information so that informed knowledge can precede events. One way of ensuring that informal communication, such as rumour, is not distorting facts, is to see that formal communication is plentiful and continual so that it anticipates and guides information passing through informal channels.

Our research in Royal Exchange and Carters illustrates some of the communication difficulties likely to arise at this time. The principal sources of information in each firm were the computer centre personnel—the programming staff and, in Royal Exchange, the O. & M. manager. These groups were poor communicators for a number of reasons. First, they were isolated in the computer centres, and therefore unaware of staff information needs until they actually visited a department or branch to prepare for the change-over. Second, they had their own problems of getting the new system to work; they were extremely busy and, even if they had been aware of the benefits of good communication, they would not have had time to devote to this. Third, because they were the innovating group, to many staff they represented 'the enemy'. We suggest that at a time of technical innovation, communication policy and procedure should not be left in the hands of technical specialists, but made the responsibility of a specially set up committee, or a neutral manager, who can devote considerable time to identifying information needs and seeing that they are met in the most efficient and advantageous way.

Management and communication Effective communication depends on people, systems and techniques. Let us consider first the communication roles and problems of the different management groups within an organisation during a period of technological change. Directors and senior management may not be actively involved in downwards communication at this time, but they will have their own communication problems. Mann and Neff (1961) point out that it is not at all uncommon for top management to decide to go ahead with a change which has been thought through and proposed by a small sub-group in the organisation, without any real depth of understanding about the long run, indirect, and marginal implications of the change process. In many firms there seem to be serious communication problems between computer centre personnel and top management. Computer specialists may make requests which a technically unsophisticated directorate cannot evaluate or even understand. It is then common to resort to the ubiquitous consultant. Mann and Neff believe that complex technological changes must be thoroughly understood by the key executives of the organisation or it may not be possible to carry these changes through to a successful conclusion. The President of the American Society of Advanced Management has said 'many managers would rather live with a problem they can't solve than accept a solution they do not understand'.

Here then is a serious and urgent communication problem for senior managers. How can they learn about the new computer technology and, even more important, understand the implications of the new information system for their own organisation? Computer courses run by manufacturers and consultants are of value, but the social and organisational consequences of electronic data processing are largely a sociological problem and there is as yet little senior management training in this area.

Middle managers of departments affected by the change have problems both as recipients of information and as communicators. The attitudes of managers to change will certainly influence the reactions of their staff, and managerial attitudes will depend to a considerable extent on what they themselves know about the change and how they view it. In Royal Exchange and Carters we saw how, before the change-over, branch managers were ignorant of what was to happen and anxious because of this. Lack of information means ambiguity and uncertainty, and the probable misunderstanding of the purpose of the change. If managers do not know what a particular change involves, how can they interpret the meaning of change to their staff or enlist their co-operation? How can they pass on to top management the apprehensions that their staff are feeling? It would seem obvious not only to consult with middle management before going ahead with technical change but also to keep them fully informed of all developments and to make them responsible for passing information on to their staff. Yet, as we have seen, our firms did not do this.

The innovating groups in the computer centres are perhaps more in need of communication expertise than anyone. Although, as we said earlier, they should not be responsible for communication policy, they will certainly be involved in it. Even if someone else organises communication procedures, they will be left with the problem of communicating their ideas to top management and, inevitably, they will be in face-to-face contact with middle management and clerks when the change-over is taking place. To most people they will seem highly ambiguous figures and what they say and how they behave will be used as a basis for evaluating the new system.

A considerable problem with this group will be giving them an understanding of the importance of human relations factors in the acceptance of change. Often their training has been in mathematics and, because a certain procedure is entirely logical to them, they expect others to regard it immediately in the same light. They may have little understanding or tolerance for the misgiving of clerks affected, for the first time, by a major change.[1]

There are many examples of how the acceptance of change has been slowed up by communication difficulties or the lack of awareness of the need for communication on the part of the innovating group. Miller (1960) cites a telephone company where there were considerable difficulties during the computer installation period between methods analysts and supervisors; and a drug company where 'the most critical period came with the dismissal of a methods analyst—a situation resulting from his personal difficulties with the other groups'. One of the British firms we visited had to get rid of a highly qualified mathematician because of his contemptuous attitude towards the staff whose work he was changing.

In Royal Exchange, although the programming staff had so recently been clerks themselves, in assuming their new mantle of technical specialist they appeared to forget the needs and problems of staff in the branches. In Carters, the programmers saw the need for communication but were forbidden to give information by senior management—this made their role both difficult and embarrassing.

Computer centre staff are therefore in a most difficult position. Good communication is vital if the new methods they are introducing are to be accepted, and they must be able to communicate their ideas both upwards and downwards. Here training would seem one answer to their communication dilemma. Training first in an understanding of the human relations problems associated with change so that they know the kinds of reactions they are likely to meet with. Training also in when and how to get their ideas across in an easily understandable way. Both managers and clerks in Royal Exchange complained to us that computer centre staff talked 'jargon' which the ordinary person could not understand.

Lastly, it must be stressed that communication during a period

[1] One British computer manufacturer is now using group dynamics and sensitivity training to give its systems analysts some insight into human relations.

of change must not be left to chance, but should be highly formalised. Communication should be the result of a well thought out and clearly defined policy in which each management group must have a specified role. An example of a communication strategy will be given in Appendix B. Miller (1960) makes the point that poor communication is by no means invariably caused by lack of good intentions and effort. He believes that we do not yet know enough about the barriers to successful communication and that it is easy to have form without substance, or an attempt without reaching the desired result.

Systems and techniques Motivating management to communicate and giving them defined communication roles is not sufficient unless attention is paid to the mechanisms for communication. These are the communications system or network that exists in the firm and the techniques used for passing information along it. Both of these may need to be modified and extended during a period of change. This is a vast subject and it is only possible to make a few points here.

One of the difficulties with the unsystematised communication at Royal Exchange was that information was sporadic. The change-over to a new system took several years, yet formal communication occurred only in the early stages and dropped off after the first branches went on the computer, being replaced by rumour. But there was need for information to be sent to staff at regular intervals throughout the change period (a monthly news letter would have been a possibility). Staff particularly wanted to know what was taking place in other branches, when their own work was scheduled for transfer, and the reason for the continual delays.

In other firms we found that although a considerable amount of official information was sometimes sent out, there was no check that this actually reached its destination. One company paid a great deal of attention to keeping departmental managers well informed and relied on them to pass the messages on to their staff. They did not do this. Mann and Neff (1961) suggest here that formal and informal channels need to be monitored regularly during a period of change in order to be certain they continue to be available and open. One reason for this monitoring is that people on the receiving end of change are often reluctant to accept disturbing information. They are worried and 'don't want to know', pretending, perhaps, that if they 'don't know', it will never happen.

This leads us to another point—that of information content. Communication lines may be good but if communication content is threatening, people will react adversely. Communication about change needs to be as reassuring as it can be in accordance with honesty. It would not be ethical to say nothing about, or play down, the adverse consequences of change, but frequently staff are worried about things that are not likely to happen and it is quite possible to provide genuine reassurance. Computers are usually associated with redundancy, therefore the first information clerks need to have is that they will not lose their jobs.

Considerable thought needs to be given to communication techniques, and if these are varied, communication is likely to be more effective. Royal Exchange relied primarily on one or two lectures from the O. & M. Manager in the early stages of computer installation, on articles in banking journals, on the circulation of written instructions and on informal contact between computer centre and branch staff. Our research has shown that this was not enough. Clerks, even if they attended, quickly forgot what had been said at a lecture, and very few had read journal articles. Yet our analysis demonstrated that those clerks who had heard about the computer from an official source were much more likely to approve of it than those who heard informally. Regular lectures for all interested staff would have helped, together with a monthly news sheet, and the opportunity for early discussions with computer centre staff and the clerks' own branch managers.

When a branch was actually in the process of changing over, clerks criticised the fact that they received 'screeds and screeds' of written information which meant nothing to them. This was a time for visual communication. The clerks wanted to see for themselves what had happened at other branches, and they would have liked to have seen and learnt about the new computer that was having such an impact on their work.

To sum up, communication is a most complex subject which, because it is to a large extent a reflection of existing social relationships, firms have great difficulty in handling successfully. A period of technical change places extra demands for speed and accuracy on communication lines yet, because it also arouses a great deal of emotion, easily causes these lines to become blocked and ineffective. Some of the points we have made in this section will help a firm to be aware of communication pitfalls, but the same moral applies to communication as to all other aspects of change. The adoption of general principles is not enough; each firm must look at its own situation, consider its objectives and analyse the communication needs of different groups of staff in terms of these objectives, including that of the least possible anxiety and stress. It must then design and implement a communication policy tailor-made to its own requirements.

CONSULTATION AND PARTICIPATION

A firm introducing a computer must consider not only its communication policies but also the extent to which it is going to consult with staff and involve them in planning and decision taking. Neither Royal Exchange or Carters did very much of this. Carters' secrecy policy made consultation impossible and extensive consultation with staff has never been part of traditional banking culture. Also there were no militant unions in either firm to insist on democratic procedures. Carters' girl clerks were not unionised and the National Union of Bank Employees, which might have taken a firm line, was not recognised by Royal Exchange. This lack of consultation was fairly typical of the other British firms we observed. In most, the change was pushed through by O. & M. specialists or accountants who were either not prepared to consult, or did not believe consultation would be of any value. We were left with the impression that, although British industry pays lip service to democracy, unless there is pressure from either trade unions or an established consultative committee, little consultation about technical change takes place. Managements prefer to introduce the change as they think best and leave their staffs to adjust to it.

In view of this absence of consultation, it seems important to examine whether there is a case for it. Is there any evidence to show that it facilitates the acceptance of change, or does it merely increase top management's difficulties by slowing up the change process and causing argument and dissention?

In general, what evidence there is suggests that consulting with staff and allowing them to participate in decisions on problems that concern them, does pay dividends, providing certain other conditions are met. Mann and Williams (1960) found in the firms they studied that allowing management to participate paid off in a minimum of inter-organisational conflict at all levels. Miller (1960) found in a public utility that 'many early problems were intensified because key supervisors and employees had not been consulted enough about the changes that were taking place; and, in a bank, that there was universal feeling that staff should have entered more actively into the planning'. Coch and French (1948), in a well-known piece of research, came to the conclusion that consultation and participation led to positive motivation towards change. Employees had an opportunity to contribute their ideas, an understanding of all the facets of the problem, a clearer definition of the objective, a sense of responsibility for the success of the decision, and a feeling of satisfaction with the course of action agreed on. Mann and Neff (1961) suggest that 'an organisation which has trained its managers and supervisors to use a management philosophy of participation is better prepared to understand and handle subordinates' increased needs for information during a period of ambiguity and uncertainty'.

So far so good. Research appears to support the hypothesis that consultation and participation facilitates the acceptance of change. But there are certain provisos to this statement. One is that staff must want to participate and feel they have something to gain from it. French (1960) followed up at a later date his early experiments in participation and decided 'if people do not feel that it is right and proper for them to influence and have a voice in a matter, sharing a problem will not result in the positive motivational effect attributed to participation'. He also found that participation produced results only when there was low resistance to change; where there was high resistance it made no difference. Mann and Neff (1961) make the point that 'felt' participation is more important than 'actual' participation. If a person feels he has been involved, the actual amount of participation may not be important.

Two conditions for successful consultation and participation then are that people must feel it is right for them to be consulted, and that they should not be too resistant to the proposed changes. If we accept these criteria, the managers of Royal Exchange and Carters should most certainly have been involved in planning the change. They wanted to be involved, had much to contribute, and most of them appreciated the need for their firms to automate office procedures. Similarly, some degree of consultation with male clerks might have prevented the drop in their morale. It is less certain whether participation would have been of value with the girl clerks. They were not particularly interestcd in work and not used to consultation. This could either have embarrassed or bored them.

Crozier (1964) suggests that people are ambivalent towards participation. On the one hand they like to control their own environment, on the other hand, if an individual accepts participation, he is bound to co-operate, and not everyone wishes to co-operate in an undesired change. There may therefore be quite rational reasons for an unwillingness to participate, even when management offers this. Also some personality types do not welcome participation. Highly authoritarian or dependent personalities may not welcome it and may prefer to be told what to do (French, 1964).

Whether consultation and participation in decision making produces favourable attitudes towards change depends then on a number of variables in the situation; whether people want to be consulted, whether they are not too opposed to the change, whether they are of a certain personality type. Here again managements must analyse their own circumstances and decide on an appropriate course of action. It does seem, however, that less damage is done if staff are offered participation, even though they do not want it, than if change is imposed on groups and individuals who feel strongly they should have been consulted.

Mann and Williams (1960) describe an example of the successful use of participation. In one firm, two years before the computer arrived, a series of meetings was organised for men at middle management levels to explain what was being attempted and to begin to interest them in the planning and execution of the change. After planning had been under way for a while, weekly meetings were held for those managers whose work would be most affected. Because managers and subordinates were notified in advance of each step and involved in the development of plans for change, conflict was kept at a minimum. Mann and Neff found that the period of change provided the firm's management with an invaluable opportunity to develop an explicit philosophy; 'we do this', not 'if something happens we would do this'.

TIMING AND PHASING THE CHANGE

The Operational Research Manager of a large British retail organisation now assimilating its third computer has told us that the most important factor influencing the rate of change in his firm is the time required to allow staff in affected departments to accept and adjust to the idea of a new system. This phasing of change is a difficult one for management who are caught between two contradictory objectives. On the one hand is the desire to press the change through as rapidly as possible in order to secure the quickest financial return on capital investment, on the other the wish to minimise stress for those employees who are introducing and are affected by the change.

Both Royal Exchange and Carters were able to give staff plenty of time to adjust, although, as we have seen, they by no means eliminated stress and anxiety in doing so. But in neither firm was this time intentional or planned for; it was solely the result of two technically unreliable computers. The management of the two firms had planned a much faster changeover, and had been led to believe that this was possible by their respective computer manufacturers. When breakdowns caused delay great pressure was exerted on computer staff to produce faster results. Mann and Williams (1960) found a similar situation in the light and power company they investigated. They report: 'target dates had been set early, on the basis of inadequate knowledge. Fears of not being able to meet the dates and failure to meet them caused psychological stress'.

In these days of greater computer reliability, time schedules for change can be planned much more accurately. Research suggests that managements will gain in the long run if they do not try to rush the change through. Mann and Neff (1961) believe that 'the speed at which a change is introduced is a key factor in determining how effective the change will be. A change is more likely to be smooth, appropriate and lasting when its introduction allows for a greater exchange of information between those engineering the change, and those being directly affected by it.' Mann (1962) also suggests that one of the reasons why office automation in insurance and public utility companies has so little effect is because there is more time available to cushion the effects of change. In our research, Royal Exchange certainly had this time as we have seen, but it did not use it.

One way of avoiding the stress arising out of rapid change is to start the preparation and planning for electronic data processing early enough. There is usually a time lag of about twelve months between the ordering of the computer and its installation, and several months will elapse after this before any work can be processed on the new machine. Provided that procedures to inform and consult with staff, and to get them used to the idea of a new system, are introduced immediately the computer is ordered, excessive stress can be avoided. This pre-planning is especially important if the computer is expected to displace staff, for running down a labour force through natural wastage takes time.

However, in order to start working for the easy acceptance and assimilation of change as soon as the computer is ordered, policies for change must have been devised even earlier than this. This means that when a computer is first being considered, management should start finding out about the problems of change and deciding what kinds of policies will prove most effective for making it acceptable in their own situation.

REWARDS

Even though change is extremely well planned and phased by a firm, some groups of staff are going to have to work very hard indeed for a period of time. In Royal Exchange and Carters we saw how the programmers worked almost day and night during the early days of computer installation. Similarly, managers and clerks in the branches and departments of the two firms had considerable overtime when their time for transfer to the new system arrived. Yet managements often seem to take all this extra work for granted, and expect clerical staff to do additional work for little or no extra reward, in a way that would be unheard of on the shop floor. This then is another policy area which will affect attitudes to change. Staff who feel they are being exploited will be less favourably disposed towards a new system than staff who believe their additional labour is both appreciated and recognised by management. Money is often a more effective way of showing appreciation than thanks.

The objective consequences of change

VARIABLES AFFECTING ATTITUDES TO CHANGE (4)

The objective consequences of change

Redundancy	*Displacement and transfer*	*Job change*
The number and kind of staff having to leave the firm.	The number and kind of staff having to move from an affected department.	What happens to particular jobs? Do they become more or less interesting; more or less status conferring, etc.

These were discussed at some length in the last chapter and all that is necessary here is to make the point that they are additional factors conditioning attitudes to change.

If redundancy occurs through the introduction of a computer, obviously this will affect attitudes towards the new system. Unless preparations are made to keep hardship to a minimum, staff will not look favourably on the change. The word 'redundancy' usually has the effect of making staff look for other jobs 'so that it doesn't happen to me', and it is the staff the firm can least afford to lose who tend to get out first. Therefore if redundancy is unavoidable, everyone must know well in advance precisely which groups will be affected. There must be no uncertainty in this most important area of employee security.

Similarly, if staff are displaced from one department and have to move to another, this will condition their attitudes. If it is not possible to show that they will gain from the move, at least every attempt must be made to see they do not lose. Again financial compensation is often an effective way of securing acceptance.

Lastly, if a job changes its nature because of technical innovation —as happened with the Royal Exchange machinists—steps can be taken to avoid discontent. If it becomes more monotonous and routine, job enlargement or job rotation are possibilities. Most important, recruitment policies need to be revised so that individuals are engaged for the work who can tolerate monotony and

An interaction model of variables affecting attitudes to change

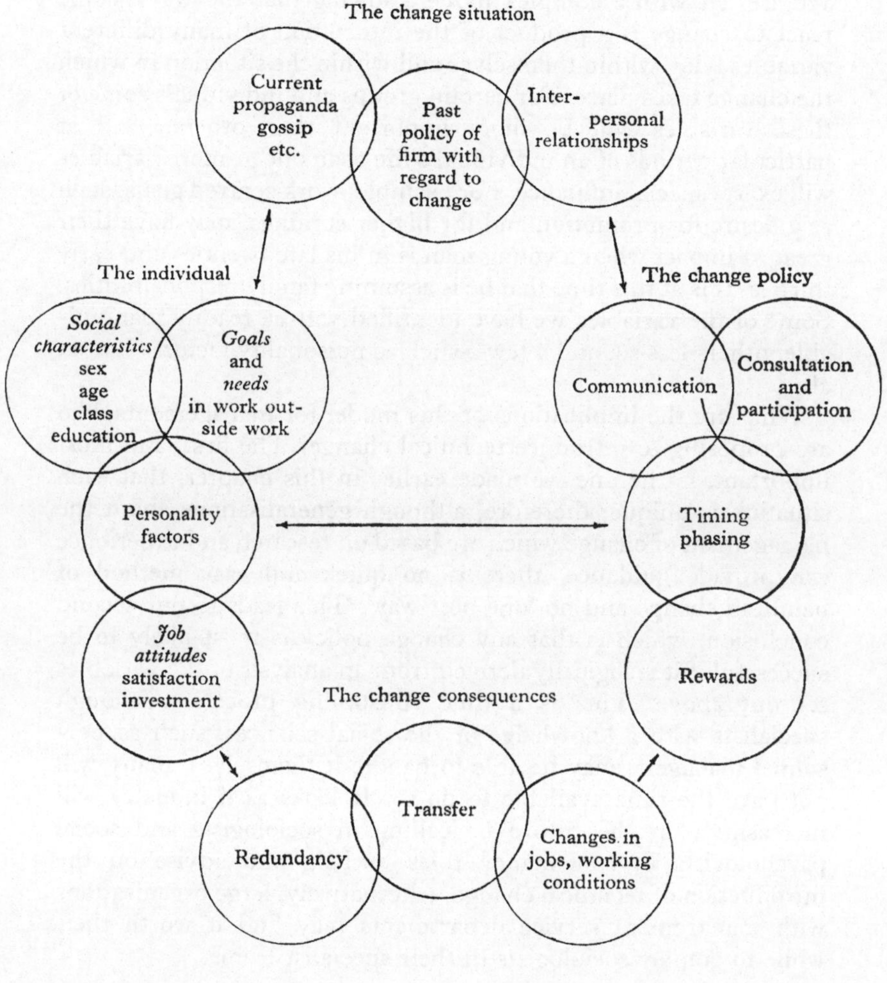

whose educational attainments do not make simple jobs inappropriate.

CONCLUSIONS

We are left with a complex model showing that the way people react to change is a product of the interaction of many different variables lying within themselves and within the situation in which the change takes place. For certain groups and individuals some of these variables will be more important than others. And at particular periods of an individual's life span one or more variables will exert especial influence. For example, work centred goals, such as a desire for promotion and for higher earnings, may have their greatest impact when a young man is in his late twenties and early thirties. It is at this time that he is assuming family responsibilities. Some of the variables we have identified will be readily manipulable, others less so, and a few—such as personality factors—not at all.

What are the implications of this model for managements who are proposing to introduce technical change? The first, and most important, is the one we made earlier in this chapter; that each situation is unique, therefore, although generalisations about the management of change which are based on research and experience can provide guidance, there is no quick and easy method of handling change and no 'one best way'. This leads to our second conclusion, which is that any change policy is most likely to be successful if it is logically derived from an analysis of the variables set out above. This is a time consuming process. Although specialists with a knowledge of the social sciences, such as personnel managers, may be able to handle it themselves, many will not have the time available to do so. It looks as if industry will increasingly in the future be calling in sociologists and social psychologists in consultancy roles to help and advise on the introduction of technical change. Alternatively, large organisations with management service departments may find it worth their while to employ sociologists in their specialist teams.

CHAPTER 11

A practical approach

In the last chapter we discussed the variables likely to influence attitudes to technical change. In this one we set out the steps necessary to identify these variables and to develop a plan of action for introducing change.

All managers when faced with a new problem carry out some analytical processes even though they are not always aware of doing this. Decisions based on what is called 'judgement' or 'intuition' or 'experience' are really the result of a complex mental process. The manager will either consciously or unconsciously try to identify the significant variables in a situation; he will consider how these variables interact with each other and, using this analysis in conjunction with knowledge usually derived from past experience, he will decide on a strategy for achieving a desired objective. Obviously a manager can never have information on all the variables operating in a particular situation, but if his preliminary analysis is so inadequately done that he misses relevant variables of major importance, then his subsequent decisions will be poor and will not lead to the attainment of his objectives.

The approach we give below should help managers both to identify relevant variables in their own situations and to think logically about how a technical development will alter the interaction of these variables.

Five steps seem essential to the successful introduction of technical change; the first three define the problem, the last two attempt to solve it. First of all, the firm's objectives need to be clearly established. Why is a new piece of equipment thought necessary at this time? What use is to be made of it, will this usage alter over time, what short and long term benefits are visualised? The firm will also need to consider what resources of time and personnel it has to achieve these objectives. Secondly, a period of preliminary research is necessary. Before the equipment is ordered, management, if it is wise, will try and obtain as much information as possible about alternative systems, technical limitations and the kinds of installation and usage problems other firms have encountered. With a computer this will involve managers in reading available literature, attending courses and paying visits to firms using computers for purposes similar to those envisaged by their own company. Thirdly, there is a need for painstaking preliminary analysis of a technical innovation's consequences. As our research data has shown, a firm introducing a computer needs to try and identify in advance the machine's technical, social and organisational consequences. At the same time, management must decide whether the nature of its business requires that particular weight be given to any one of these areas of change. Unless this analysis is undertaken, planning cannot be done from an informed position; the extent, nature, and component parts of the change problem will not have been clearly defined.

Once the three steps listed above have been completed, it is possible to start thinking in terms of action. The fourth and vital step is devising a plan for introducing the change. Here the most fruitful approach will be to design a number of plans all calculated to achieve the firm's objectives with a minimum of social and organisational disruption. After these have been discussed with interested parties a final plan can be selected, the criteria for choice being maximum benefit for least social and economic cost. Fifth and last the chosen plan must be put into operation. In order to allow for unanticipated contingencies in a dynamic situation this plan will need to be flexible enough to allow for changes to be introduced if necessary. This means that a phased plan is likely to prove the most successful. Alterations can then be made to parts without affecting the structure of the whole.

Let us now examine each of these steps in terms of the experiences of our case study firms.

Objectives

These will differ from firm to firm but it seems reasonable to suggest that most firms will favour a strategy for change which aims at maintaining staff morale during the change period, at reducing the stress and conflict normally associated with any major change and at facilitating the acceptance of change. These human relation goals will, of course, be operating side by side with economic objectives, such as a reasonably fast return on financial investment. In some firms economic goals will assume the greatest importance; in others, human relation goals will have equal or even greater influence; It is unrealistic to expect a firm to give overriding consideration to human relations objectives if it is going to bankrupt itself in doing so. However, some firms, like banks and insurance companies, where there are few pressures for quick profits, should be able to pay a lot of attention to staff needs provided they are aware of what these are and can be persuaded that to do this is worth while. Other firms, such as those which have difficulty in securing and retaining staff, may have no option but to do the same; if they do not include human relations considerations in their objectives they will have serious personnel problems. Our research material has shown that both Royal Exchange and Carters could have paid more attention to human relations. Their failure to do so was due not to a disregard for these but to a lack of awareness of the major effect a proposed technical innovation can have on staff attitudes.

Research into alternatives

The Royal Exchange O. & M. Department did do some thorough research before embarking on the new system, but in Carters, and in the other British firms we visited, we were surprised at how little preliminary investigation was undertaken to ascertain the kinds of problems likely to arise when introducing a computer for the first time. Managers showed little evidence of having read journal articles on computer installation, and few had bothered to visit other firms to see how they had fared and the kind of problems they had encountered. There was a tendency to play the change process by ear, to leave it entirely to technical specialists, or to assume that introducing a computer was as simple as bringing in some elementary piece of mechanisation such as ledger posting machines. Most firms paid considerable attention to possible technical problems, but ignored the social and organisational impact of electronic data processing. They therefore neglected highly important variables in the change situation.

Analysis of consequences

In order to predict the consequences of technical change, analysis needs to cover three areas; work, organisation, and behaviour. The analysis of work involves estimating the number and nature of the jobs that will be altered by the new system; identifying those jobs that will be eliminated altogether and those that will be partially or completely changed. This is not easy to do with any exactness before the computer arrives, but it must be attempted. A competent systems analyst with a knowledge of computers will be required to undertake it.

The analysis of organisational change will be much wider than the analysis of work, covering changes in information procedures, decision making processes, and departmental and management structure. Major change in these areas will, of course, only occur if the computer is used for integrated data processing, but a lower level of computer usage will still have some organisational impact, particularly on information and decision procedures. Again this analysis is not easy to do in advance of change, and there are not yet many managements or consultants in this country experienced in undertaking it. But Shultz and Whisler (1962) claim on the basis of their American experience that, in a specific application, it is quite possible to forecast the location and nature or organisation changes, once the new patterns of information flow have been determined.

The analysis of the behavioural response to automation—predicting how employees will react to the idea and the reality of change, is really a job for a social scientist, although Lupton (1965) suggests that managers can attempt it themselves providing they use the methods of the social scientist as a tool for analysis. If the social analyst wishes to anticipate the behavioural responses to change of a group of employees he will need to consider the variables described in the last chapter. He must be aware of the social characteristics, personalities, job attitudes and goals and needs of those employees about to be faced with change, and he must be informed on matters such as past change policy, existing interpersonal relationships, and current propaganda, which form part of the change situation. The methods for acquiring this kind of information are well documented in the research literature of the social sciences.[1]

[1] See, for example, Madge (1953) or Festinger and Katz (1953).

Reasonably accurate prediction of how an envisaged technical change will affect work, organisation and attitudes makes it possible for management to evolve change strategies which are likely to be successful. It now becomes possible to work from an informed position and to replace supposition and guess-work with knowledge.

Planning

Planning must cover every aspect of the change process and it is too complex a subject to be discussed in any detail here. However, those areas most likely to influence staff acceptance will be communication and consultation, manpower planning and the timing and phasing of change procedures.

Communication will involve four decisions; who to tell, when to tell, what to tell and how to tell. A further decision must be taken on which groups will require to be consulted and involved in the planning process. A brief case study on communication is to be found in Appendix B.

Manpower planning follows on from the analysis of work. To what extent and in which areas will the existing labour force need to be increased or run down to meet new requirements? How soon must this process start? Will staff need to be transferred to other departments? Are suitable vacancies available for them, and how much re-training will be involved? What new specialist staff will be required to run and maintain the new equipment, and how can they best be obtained and trained? These are all questions that will need to be answered by a firm's personnel department.

Phasing and timing will include technical, economic and social factors and the following questions will be relevant to these. How long will it take to run in a new piece of equipment? What is the maximum allowable time before a new system must be operationally effective and paying for itself? How much time is required to secure easy acceptance and assimilation by affected staff? How long will it take to train staff and for them to become competent in operating the new system?

Execution

This is when any deficiencies in the analysis and planning stages show up. Planning is abortive unless the plans are successfully introduced and carried through. As we pointed out earlier, plans should never be so rigid that they cannot be altered if circumstances suddenly change.

A factor crucial to the successful execution of plans for change is a decision on who is going to be responsible for their implementation. Shall the planners also be the executors? If we accept that it is an advantage to involve in planning those managers and supervisors whose departments will be altered by the change, the answer to this will be 'yes'. A strong case can also be made out, however, for the introduction of a new managerial role at this time, that of 'change agent'. The change agent functions as a communication link between a number of different social systems (Rogers, 1962). In the case of computer introduction these will be the computer centre, the affected departments and the firm's senior administrators. He may take responsibility for checking that plans are being implemented as designed, and for spotting areas where plans may have to be adjusted or re-formulated to meet unanticipated contingencies. The personnel manager or a member of the management team with personnel experience could assume this role, or an outside consultant or a specially set up committee be asked to undertake it. Normally it should not be a member of the computer team. Their operational responsibilities make it difficult for them to maintain a detached point of view, and they will invariably be pressed for time through their technical commitments.

The analytical approach discussed above is linked to the introduction of a computer into a firm which has previously handled its clerical procedures non-electronically. However, it must be emphasised that this kind of research, analysis and planning will be a continuous process. When a firm decides on E.D.P. it is usually embarking on a long term process of progressive change. As more and more procedures are transferred to the computer, and the firm increases the size of its computer installations, so more complex organisational changes will result from the new technology and will need to be identified and planned for in advance. Knowledge based on previous experience will be a help here, but, because no two situations are identical, it is no substitute for the painstaking hard work of identifying the consequences and requirements of a constantly changing system. The question must then be asked, is it really worth while to take all this trouble rather than to muddle on as so many firms do? We suggest that this is something each firm must decide for itself. Any major change involves considerable adjustment on the part of the social system in which the change takes place. We believe that with comprehensive analysis and planning, these adjustments can be anticipated, intended and recognised by the participants. There is, therefore, complete control of events, and management is not left desperately trying to devise ad hoc measures to deal with one unexpected crisis after another, as so often happens at the moment.

APPENDIX A

Methodological problems

Methodologically and operationally this project has presented us with a number of problems. Its international nature imposed certain constraints on the participating research teams who, to some extent at least, had to conform to a common research plan. In the early stages of the research, when the project was being designed and many countries were involved, this need for agreement on future action caused difficulties, although as countries dropped out and the number contracted to an eventual three—France, Sweden and Britain—these co-ordination problems became less. A more serious handicap arose out of the decision to make the project an on-going study of change. At the start of the research it was optimistically assumed that any transfer of work to E.D.P. could be completed within twelve to eighteen months of a computer's arrival in a firm. We had no conception of the kinds of delays that were going to occur before the project could be completed.

These delays, it should be stressed, were not due to any lack of haste on the part of the two firms. They were anxious to move to E.D.P. as speedily as possible and the holdups were usually due to events outside their control. In the bank, the computer arrived two months after it was expected; technical difficulties made the period of parallel running an extensive one, and the very delayed delivery of a line-printer held up for many months the transfer of work of the last two branches. Three years after the computer had been installed the work of only three out of the five branches was being processed on the computer. Carters' machine arrived on time, but this firm encountered even more severe technical problems than the bank and, in this firm too, three years after installation only two-thirds of the scheduled work had been transferred. From a research point of view these delays were extremely frustrating. They made precise planning of the project an impossibility, and delayed the publication of results which it seemed important to produce as early as possible if the work was to be of use to firms contemplating E.D.P. In addition, between 1961 and 1964 tremendous technical developments in computer design had taken place: we had to face the problem that our study of what were, in fact, first generation computers, might from the practical point of view be out of date. It was fortunate for our morale that we were not alone in our difficulties; research teams in other countries were experiencing similar delays and for identical reasons.

Not only was the research held up for long periods, but the collection of data was by no means easy. Firms feel vulnerable and exposed when they are undergoing major change and are not anxious to be observed too closely by outsiders, no matter how disinterested these may be. Both the bank and Carters were worried that the pre-change interviewing programme might make their staff anxious about office automation and cause them to resist the introduction of the new computerised system.

Carters did not wish its staff to know that the firm was even considering a computer and so they requested us not to ask their clerks any questions on the likely impact of a computer within the firm—we were permitted only to ask questions on the effect of automation of clerical work in general.[1] The bank did not impose any censorship on the questionnaire but a number of branch managers were at first hostile to the idea of research and interviewing. Banks have a tradition of secrecy and security, managers tended to feel that they should not be exposed to investigation and they were worried that their clerks might be 'upset'. Although interviewing was permitted in all five branches, permission was often given reluctantly and in one instance a branch manager had a prior meeting with his staff and warned them to be careful what they said to the interviewer. The research team had to proceed very cautiously indeed, as they felt that one indiscreet move would bring interviewing to an abrupt halt.

Interviewing itself was not easy. Many clerks, brought up in an atmosphere of confidentiality, felt that their attitudes to work were their own affair and should not be disclosed to an outsider; others could not see the relevance of the survey although this was always carefully explained before an interview was started. Fortunately

[1] This request had also been made by management in the early A.O.C. study.

for our research objectives they were often unaware of the kind of information we were particularly interested in, and so would suppress details of bank organisation while talking at length on their attitudes to work and problems associated with the bank clerk's role and status. In Carters, interviewing was not made difficult because of an unwillingness to talk but the respondents' extreme youth and job inexperience meant that they had often little to say. Work was of small importance in their lives and questions about the content of their jobs and the likely effect of automation aroused little interest.

Although eventually most of these problems were overcome and the project was brought to a reasonably successful conclusion, it proved a most difficult piece of research to handle, both operationally and from the public relations angle. The two investigators had both had considerable experience in interviewing shop floor groups such as steel workers, dockers and miners. They came to believe, perhaps subjectively, that the manual worker is a far more co-operative research subject than the white-collar worker, and particularly the white-collar worker who comes from a closed and secretive community such as that associated with banking.

APPENDIX B

Rurditch City Council—a communications strategy

(This case study is fictitious, although based on actual experience)

Proposal

The Rurditch City Council is shortly to install its first computer in City Hall. This machine will, initially, take over the work of four departments—Rates, Wages and Salaries, Purchasing and Bills and Claims. All these clerical operations have previously been performed manually. Although the kind of work carried out by the clerks in the four departments will be drastically altered, there will be no redundancy. Any required reduction in staff numbers will be achieved through natural wastage.

Principal objective

To carry out the transition from one system of work to another as smoothly and quickly as possible, while at the same time causing staff the minimum stress and anxiety.

The City Treasurer appreciates that the first step towards the achievement of this objective is the prevention of alarming rumour, therefore before taking any action he lets all his staff know that a computer is being considered, but that nothing will happen for some considerable time. He tells them that they will be kept fully informed of all developments and that there is not the slightest danger of redundancy.

Research

He appreciates that it is essential to secure prior knowledge of the kind of problems other local authorities have experienced when introducing computers. As his Authority has an Organisation and Methods department he asks the O. & M. Manager, who is also an accountant, to take reponsibility for securing information on likely technical problems. The Staff Manager is asked to do the same for social and manpower problems, and also to commence some preliminary talks with representatives of the employees' union—N.A.L.G.O.—the National Association of Local Government Officers.

In order to become knowledgeable himself, the City Treasurer obtains a sample of available literature on the technical, social and organisational problems of computer installation and reads this. He also goes on a computer appreciation course with his deputy and visits a number of comparable computer installations.

Analysis

WORK

A systems analyst experienced in electronic data processing is engaged and made responsible for ascertaining which jobs will be changed, how they will be changed, and the number of staff likely to be affected. In his report he indicates that, although there will be no redundancy, some staff will be displaced from the four affected departments and these will have to be absorbed elsewhere. He describes how the jobs of clerks and supervisors will change and the implications of this for training policy and procedures, and he estimates the number of staff that will be required with the new system. He points out the areas in which the jobs of the four departmental managers will be altered.

ORGANISATION

The City Treasurer believes that the organisation and structure of the Council's administrative departments will not be changed by the first computer developments, but he is aware that this may be a problem as electronic data processing is extended. He therefore proposes to watch future developments carefully and to take steps to identify in advance the organisational consequences of policy decisions on office automation.

BEHAVIOUR

He hopes that his staff will be prepared to accept the new system but believes that many will not welcome it and that some may be very apprehensive. He asks a social scientist to come and interview his staff so that he may have some precise information on their attitudes.

In his report the social scientist makes the following points. Few of the staff have any specific knowledge about computers and their effects but the press, television and friends in commercial organisations with computers have given them some pre-conceived ideas. It appears that,

(a) Departmental managers are worried about their own status and autonomy; about the removal of some of their work and the reduction in staff numbers under their control.
(b) Female clerks associate a computer with increased routinisation of work and a decline in work interest. Well qualified girls are particularly anxious about this.
(c) Male clerks are disturbed in case promotion opportunities decrease as a result of the computer. They wonder if a knowledge of computer technology will be necessary for upgrading in the future.

Planning

The City Treasurer's next step is to set up a team to devise and implement the change policy. This team consists of the four departmental managers whose work will be affected, the Organisation and Methods Manager, the Staff Manager and the City Treasurer himself.

The change team are asked first to formulate a policy for communication and consultation. The O. & M. Manager tells them that the computer has now been ordered and will arrive in twelve months time. Six months will be spent finalising programs. The four departments will then go on in turn at about three monthly intervals. This means that there will be a gap of eighteen months before the work of the first department is transferred.

The change team starts its work by considering the communication problems it must provide for. It decides:

1. There will be upward communication difficulties with senior staff in the Council Offices who may not fully understand what is going on and may be disturbed at technical developments which are quite new to them. The Council too will want to know what is taking place.

2. There will be a need for sideways communication. Managers whose departments will not be immediately affected will still want to know what is going on.
3. There will be a need for communication downwards. Affected staff will want a great deal of information. Non-affected staff will require some information.
4. Provision will have to be made for sustained communication to different parts of the organisation over a long period of time. The change-over is scheduled to take 12 months; it may take longer if unanticipated difficulties arise.
5. Decisions must be taken on *who* shall be told, *when* they shall be told, *what* they shall be told and *how* they shall be told. The preliminary social analysis has shown that different sex, departmental and occupational groups are likely to react in different ways. This means that a blanket communication policy is unlikely to be successful. Communication must be designed to meet the particular information needs of the different groups

The change team discusses how the effectiveness of communication can be ensured. It decides that measures must be taken to see that information reaches its destination rapidly and undistorted, that communication channels work smoothly and do not get blocked, that staff get all the information they need and that they get it continuously and not just from time to time. It is particularly anxious to establish and maintain effective two way channels of communication so that information can flow freely up and down the communication lines. The change team also decides that communication is not enough, there must also be consultation with all staff in the affected departments and with trade union representatives.

The team's first act is to start immediately providing some official information for all staff. The Staff Manager is asked to arrange meetings in each department at which automation plans can be fully discussed with a member of the O. & M. Department. He is also asked to provide a regular news-sheet, to be circulated to all staff, indicating the progress of these plans and the reasons for any delay.

The computer manufacturer is asked to put on data processing training courses, first for senior staff and departmental managers, later for all interested staff.

The team agrees that, as the system gets under way, arrangements must be made for all staff in the four affected departments to visit the computer centre and, in turn, to spend a day or so there. As soon as the first department is on the computer, staff from the three other departments scheduled for transfer will be encouraged to visit there and observe the new work system.

The change team now turns to its own problems and decides that it will meet at fortnightly intervals until the change-over is completed. The first meetings will concentrate on discussing the departmental managers' problems and anxieties, and obtaining their opinions on the best methods for carrying out the change. The City Treasurer is determined to do all he can to arouse the departmental managers' enthusiasm for the new system. He appreciates that it is the departmental managers who will have the greatest influence on staff reactions.

At subsequent meetings the following policies and procedures are decided on.

THE COMMUNICATIONS ROLE OF THE DEPARTMENTAL MANAGER

Each Departmental Manager will be made responsible for communication within his own department, but asked to operate within the framework of a common policy. This policy shall include:

1. The designation of a new role for one member of the staff in each department—that of computer information officer. A major part of this role will be continual liaison with staff in the computer centre.
2. Regular meetings between Departmental Managers and members of their own staff to discuss the implications of the computer and make plans to handle problems. The Manager must remember that different sections of his staff have different anxieties. Girls fear increased routine, men, decreased promotion opportunities.
3. Each Departmental Manager will report back on his discussions with his staff to the change committee. Problems common to all departments can be sorted out here.

COMMUNICATION REQUIREMENTS

Who to tell All employees will want some information about the Authority's automation plans, but clerks whose jobs are to be

directly affected will require highly specific information. Their question will be 'How is the change going to affect me?'
What to tell Each occupational group must be given the kind of information that the preliminary social analysis showed it most requires. For example, if a group is worried about something that is not going to happen, such as redundancy, they must be shown that their fears are without foundation. If, however, they are worried about something that is going to happen—as the girl clerks are about increased routinisation of work—then communication alone will not help. Information must be given on what is going to happen but joined with assurances that there will be compensation, i.e. job enlargement and job rotation.
How to tell A number of different communication techniques will need to be used to get the information across. These must include verbal, visual and written media. Feedback devices will need to be introduced to check that messages are arriving at their destination undistorted. A 'communications inspector' could be of great assistance here. He would move between the computer centre and the various departments, checking that communication lines are open and working and that informal communication is not replacing formal as the principal information mechanism.

Problems

This communications strategy should be intensive and sustained enough to anticipate and forestall rumour. However, the change team are aware that in achieving good communication they may be introducing additional problems. Firstly, staff will know in advance the areas of change which are likely to prove disadvantageous to them as individuals. They may try to take evasive action. Secondly, as we saw in Chapter 10, good communication brings conflict into the open, and this may make it more difficult to achieve a co-ordinated plan of action.

BIBLIOGRAPHY

ARGYRIS, C., 'The Integration of the Individual and the Organisation,' in G. B. Strother (ed.), *Social Science Approaches to Business Behaviour*, London: Tavistock Publications, 1962, 57–98.

ARGYRIS, C., 'Human Relations in a Bank', *Harvard Business Review*, 1954, **32**, 63–72.

BAKER, ELIZABETH F., *Technology and Woman's Work*, New York: Columbia University Press, 1964.

BANKS, J. A., *Industrial Participation*. Liverpool University Press, 1963.

BANKS, OLIVE, *The Attitudes of Steel Workers to Technical Change*, Liverpool University Press, 1960.

BARNARD, C., *Organisation and Management*, Cambridge, Mass.: Harvard University Press, 1952.

BEER, STAFFORD, 'Love and the Computer', *The Manager*, October 1963, 33–35.

BLAU, P. and SCOTT, W., *Formal Organisations*, London: Routledge & Kegan Paul, 1963.

BLAUNER, R., *Alienation and Freedom*, Chicago: The University of Chicago Press, 1964.

BURNS, T. and STALKER, G. M., *The Management of Innovation*, London: Tavistock Publications, 1961.

CLIPPINGER, R. F., 'Economics of the Digital Computer'. *Harvard Business Review*, 1955, **33**, 77–88.

COCH, L. and FRENCH, J. R. P., 'Overcoming resistance to change', In D. Cartwright and A. Zander (eds.), *Group Dynamics: Research and Theory* (2nd ed.), Evanston, Ill.: Row Peterson, 1960.

COOK, P. H., 'An Examination of the Notion of Communication in Industry,' *Occupational Psychology*, 1951, **25**, 1–14.

CROOME, H., *Human problems of Innovation*, London: D.S.I.R., 1960.

CROZIER, M., *The Bureaucratic Phenomenon*, London: Tavistock Publications, 1964.

DALE, J. R., *The Clerk in Industry*, Liverpool University Press, 1962.

DEWEY, J., *How We Think*, New York: D. C. Heath & Co., 1910.

FAUNCE, W., 'Social Stratification and Attitudes Towards Change in Job Content,' *Social Forces*, 1960, **39**, 140–148.

FAUNCE, W. HARDIN, E. and JACOBSON, E., 'Automation and the

Employee,' *The Annals of the American Academy of Political and Social Science*, 1962, **340**, 60–68.

FESTINGER, L. and KATZ, D., *Research Methods in the Behavioural Sciences*, New York: The Dryden Press, 1953.

FRENCH, J. R. P., 'Laboratory and field studies of power,' in R. L. Kahn and E. Bolding (eds.), *Power and Conflict in Organisations*, London: Tavistock Publications, 1964.

FRENCH, J. R. P., ISRAEL, J. and AAS, D., 'An Experiment on Participation in a Norwegian Factory: Interpersonal Dimensions of Decision-making,' *Human Relations*, 1960, **13**, 3–19.

GALLAGHER, J. D., *Management Information Systems and the Computer*, New York: American Management Association, 1961.

GERTH, H. and WRIGHT MILLS C., *From Max Weber: Essays in Sociology*, London: Routledge & Kegan Paul, 1948.

GINSBERG, B. M., 'The Impact of Automation on Wage and Salary Administration,' *I. L. R. Research*, 1963, **9**, 3–9.

GOLDTHORPE, J. H. and LOCKWOOD, D., 'Affluence and the British Class Structure,' *The Sociological Review*, 1963, **11**, 133–163.

GREENSMITH, D. S., 'Creating a Comprehensive Data Processing System,' *Data Processing*, 1959, **1**, 132–136.

HAMILTON, R., 'The Income Difference between Skilled and White Collar Workers,' *British Journal of Sociology*, 1963, **14**, 363–373.

HARDIN, E., 'Computer Automation, Work Environment and Employee Satisfaction,' *Industrial Labour Relations Review*, 1960a, **13**, 559–567.

HARDIN, E., 'The Reactions of Employees to Office Automation,' *Monthly Labour Review*, 1960b, **83**, 925–932.

HART, BEN, 'Why Not a Second Industrial Revolution?', *Association of Teachers of Management Bulletin*, March 1962.

HOOS, IDA, 'The Impact of Automation on Office Workers,' *International Labour Review*, 1960, **82**, 363–388.

JACOBSON, E., TRUMBO, D., CHEEK, G. and NANGLE, J., 'Employee Attitudes towards Technical Change in a Medium Sized Insurance Company, *Journal of Applied Psychology*, 1959, **6**, 349–354.

KAHN, R. L. and BOLDING, E. (eds.), *Power and Conflict in Organisations*, London: Tavistock Press, 1964.

KUSHNER, A., 'People and Computers,' *Personnel*, Jan–Feb., 1963, 27–34.

LOCKWOOD, D., *The Blackcoated Worker*, London: Allen & Unwin, 1958.

LUPTON, T., 'The Practical Analysis of Change in Organisations,' *Journal of Management Studies*, 1965, **2**, 218–227.

MCGREGOR, D., *The Human Side of Enterprise*, New York: McGraw-Hill, 1960.

MCMURRAY, R., 'Recruitment, Dependency and Morale in the Banking Industry, *Administrative Science Quarterly*, 1958, **3**, 87–106.

MADGE, J., *The Tools of Social Science*, London: Longmans, 1953.

MANN, F., 'Psychological and Organisational Impacts of Technical Change,' in Dunlop, J. T. (ed.), *Automation and Technological Change*, Englewood Cliffs, N.J.: Prentice Hall, 1962.

MANN, F. C. and HOFFMAN, L. R., *Automation and the Worker*, New York: Holt, 1960.

MANN, F. C. and NEFF, F. W., *Managing Major Change in Organisations*. Report for the Foundation for Research on Human Behaviour, Ann Arbor, Mich.: Braun & Brumfield Inc., 1961.

MANN, F. C. and WILLIAMS, L. K., 'Organisational Impact of White Collar Automation,' *U.S. Industrial Relations Research Association Annual Proceedings*. Publication No. 22, 1958, 59–69.

MANN, F. C. and WILLIAMS, L. K., 'Observations on the Dynamics of a Change to Electronic Data Processing Equipment,' *Administrative Science Quarterly*, 1960, **5**, 217–256.

MANN, F. C. and WILLIAMS, L. K., 'Some Effects of the Changing Work Environment in the Office', *Journal of Social Issues*, 1962, **18**, 90–101.

MARCH, J. and SIMON, H., *Organisations*, New York: John Wiley & Sons, 1958.

MILLER, B., *Gaining Acceptance for Major Methods Changes*, Research Study No. 44, New York: American Management Association, 1960.

MILLS, C. WRIGHT, *White collar*, New York: Oxford University Press, 1951.

MORENCO, CLAUDINE, 'Gradualism, Apathy and Suspicion in a French Bank,' in Scott W. H. (ed.), *Office Automation: Administrative and Human Problems*, Paris: O.E.C.D., 1965, 31–55.

MUMFORD, E., *Living with a Computer*, London: Institute of Personnel Management, 1964.

MUMFORD, E. and WARD, T. B., 'How the Computer Changes Management', *New Society*, 1965, **6**, 6–10.

MURDOCH, A. A. and DALE, J. R., '*The Clerical Function*', London: Pitman, 1961.

PARKER, S. R., 'Work and Non-work in Three Occupations', *The Sociological Review*, 1965, **13**, 65–75.

ROGERS, E. M., *The Diffusion of Innovations*, New York: Free Press of Glencoe, 1962.

SCOTT, W. H., *Office Automation and the Non-manual Worker*, Paris: O.E.C.D., 1962.

SCOTT, W. H., *Office Automation: Administrative and Human Problems*, Paris: O.E.C.D., 1965.

SCOTT, W. H., BANKS, J. A., HALSEY, A. H. and LUPTON, T., *Technical Change and Industrial Relations*, Liverpool University Press, 1956.

SELZNICK, P., *Leadership in Administration*, Evanston, Ill.: Row Peterson, 1957.

SIEGMAN, J. and KARSH, B., 'Some Organisational Correlates of White Collar Automation,' *Sociological Inquiry*, 1962, **32**, 108–166.

SIMON, H. A., *The New Science of Management Decision*, New York: Harper Bros., 1960.

SPENCER, P. and SOFER, C. 'Organisational Change and its Management', *The Journal of Management Studies*, 1964, **1**, 26–47.

STROTHER, G. B. (ed.), *Social Science Approaches to Business Behaviour*, London: Tavistock Publications, 1962.

TRIST, E. L., HIGGIN, G. W., MURRAY, H. and POLLOCK, A. B., *Organisational Choice*, London: Tavistock Publications, 1963.

TRUMBO, D., 'Individual and Group Correlates of Attitudes towards Work-related Change', *Journal of Applied Psychology*, **45**, 338–344.

VITELES, M. S., *Industrial Psychology*, New York: Norton, 1932.

VROOM, V. H., *Work and Motivation*, New York: Wiley, 1964.

WALKER, C. R. and GUEST, R., *Man on the Assembly Line*, Cambridge, Mass: Harvard University Press, 1952.

WHISLER, T. L. and SHULTZ, G. P., 'Automation and the Management Process', *The Annals of the American Academy of Political and Social Science*, 1962, **340**, 81–90.

WOODWARD, JOAN, *Industrial Organisation: Theory and Practice*, London: Oxford University Press, 1965.

Index

INDEX